"Unless you try to do something beyond what you have already mastered, you will never grow."
— Ralph Waldo Emerson

ONVIEW BOOKS
**SAFE MICROSCOPIC TECHNIQUES
FOR AMATEURS**
Slide Mounting

Published by Onview.net Ltd
2014

Registered Office:
Frilford Mead, Kingston Road, Frilford. Abingdon.
Oxfordshire. OX13 5NX England

www.onview.net

Copyright © Walter Dioni and Alicia Dioni
All rights reserved.
All proceeds for sales go to Walter's family in Mexico.
Visit: www.microscopy-uk.org.uk/wd/
The moral right of the author has been asserted.
Many thanks to David Walker for proof-reading this book and for
Mol Smith in compiling and editing the articles.

This book is sold subject to the condition that it shall not, by way of trade or otherwise, be lent, re-sold, hired out, or otherwise circulated without the publisher's prior consent in any form or binding or cover other than that in which it is published and without a similar condition including this condition being imposed on the subsequent purchaser.

First Published 2014 by (Onview Books) Onview.net Ltd.

**A CIP catalogue record for this book is
Available. See cover for ISBN.**

SAFE MICROSCOPIC TECHNIQUES FOR AMATEURS
Slide Mounting

by Walter Dioni
2014

Introduction and preface to this work.

From 2002 to 2011 Walter Dioni, living in Mexico, has shared over 60 often heavily illustrated articles in Micscape Magazine (www.microscopy-uk.org.uk) on a wide range of topics. They illustrate the breadth of his interests, his love of microscopic life and encouragement to others to use the microscope as a study tool. English is not Walter's first language, so it's especially impressive to witness him share technical topics in such a clear manner. All credit to him to sustain such effort in an attempt to share his knowledge freely with people around the world.

Since the 1990s, it has become increasingly difficult to obtain the types of chemicals used to preserve, treat, and mount specimens of microscopic life onto specimen slides for long term study. The increasing pressure of society to protect its young people from any form of risk, and the increasing reluctance of shipping services to handle anything deemed as slightly dangerous, has undermined enthusiast microscopy as a practical past-time. Other pressures are prevalent with the increasing attraction in young people to spend their spare time playing computer games and immersing themselves into virtual digital worlds.

Walter's work here seeks to re-enable specimen treatment and mounting through the use of chemicals and materials easy to source locally and considered safer than the more volatile but extremely efficient chemicals used in the past. The work will prove useful for expert, novice, and practiced enthusiast where they might wish to create slides of their own. Note: other Micscape authors and contributors have added to Walter's work and also included where appropriate in this book or references are provided to online articles.

At the time of writing (May 21st 2014), Walter is very ill in hospital in Mexico with lung cancer. Aged eighty-four, the prognosis for survival is very poor. This news prompted the creation of this book by the microscopy-uk and Micscape founders with an aim and hope to preserve the memory as well as the knowledge of a very intelligent and compassionate human-being. All proceeds obtained by the sale of this book will be sent to his surviving children in Mexico by way of his eldest daughter, Alicia.

Mol Smith (co-founder of Micscape Magazine)

www.microscopy-uk.org.uk

CONTENTS

Chapter 1- Introduction – Liquid Media — 6

Chapter 2 - Solidifying Media — 26

Chapter 3 - The mixed formulae — 44

Chapter 3a - Formulae Derived From Fructose FG—Fructose-Glycerol Medium — 46

Chapter 3b - PVA-lactic Acid And PVA-glycerol Mountants — 50

Chapter 3c - The Mixed Formulae - Gum Arabic Media — 58

Chapter 4 - The Glycerin Jellies — 72

Chapter 5 - Ten Years After. — 82

Chapter 6 - Finale — 98

Editor's Note

As editor, I acknowledge the somewhat lack of a proper indexing section for this work, which is due to combining a number of web based articles into a coherent work. Unfortunately, the method chosen and the resources at our disposal did not allow for a more detailed contents and indexing system to be included. We also wished to complete the work within a very short time-frame as the author is very ill and it was our wish for him to see the book in print as soon as possible.

Note on transcription of web articles to printed form: Transfer of 640x480 pixel images in the earlier web articles to high dpi print format inevitably causes some quality loss and the reader should bear this in mind. Also monochrome versions of this book lose the colour information but does reduce costs considerably.

The hyperlinking to other material in the original web format articles are also lost in the printed formats. Readers are advised to visit the full web suite of articles at the link above to follow any references. I do hope this does not detract from the highly useful content.

Mol Smith.

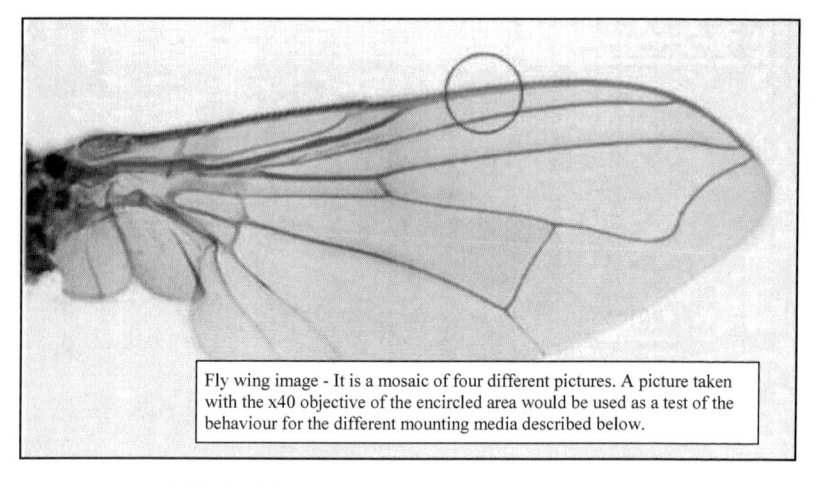

Fly wing image - It is a mosaic of four different pictures. A picture taken with the x40 objective of the encircled area would be used as a test of the behaviour for the different mounting media described below.

MOUNTING MICROSCOPIC SUBJECTS.
Chapter 1 – Introduction – Liquid Media

Introduction

In a research histology laboratory, or a pathology laboratory, mounting is the last procedure in the series that ends with a permanent histological preparation on the table, well after the:

1) fixing
2) paraffin embedding
3) sectioning
4) staining
5) dehydrating
6) clearing operations

Leaving aside some exceptions, amateurs rarely engage in research that needs long and complicated histological techniques. Most of their work is made on live material. Though at times, many of them may need or want to preserve some materials for future study, or to make a comparative collection of samples, or to see cytological detail such as a cell nucleus.

After some preliminary treatments they would like to mount the objects or organisms as semi-permanent or even permanent preparations.

Normally they manipulate whole organisms or parts dissected from them and many times microscopists mount their critters without staining. Some subjects can even be mounted without any previous manipulation at all, especially if they are dry objects. In this series on microscopic techniques, this makes the study of mounting media easier and useful, reversing what would be the usual presentation schedule.

Canada balsam.- The standard mountant for histology, and also for taxonomy, be it zoological or botanical, is Canada balsam, a now scarce and very expensive natural resin. This is prepared by collecting the resin exuded by *Abies balsamica* (the "balsam fir") and diluting it in solvents (many of which are now considered toxic e.g. xylene).

From experience to date, Canada balsam mounted preparations last over a century. As Canada balsam does not mix with water, mounting in it implies the use of a sequence of dehydration, starting with low grade alcohols, followed by high grade alcohols, absolute alcohol, mixed clearing agents plus alcohol, clearing agents, clearing agents mixed with xylene, pure xylene, and balsam dissolved in xylene, toluene or benzene could be used instead of xylene.

But all three solvents are equally toxic and dangerous: xylene, toluene, and benzene. The development of some synthetic media as substitutes for balsam don't solve the problem, they are proprietary trade marks, equally expensive, that need the same steps, and use the same (toxic) solvents. There are less toxic and less dangerous proprietary substitutes but they are expensive.

Alternatives? Amateurs normally desire an easier way to have their critters mounted and normally don't need a '100 years proof' mounting media. They need to find, easy to use, non-toxic, inexpensive, and reasonably long-lasting media (perhaps months, perhaps a few years) that assures a clear view with good contrast of the morphological traits he (or she) is searching for.

There exist several of these mounting media. A few of them are resinous, several are aqueous. But normally only one is extensively cited in an amateur's bibliography. It is glycerin jelly, a very useful option, but not the only or most easy one to use. I will review those media, selecting only the non-toxic ones, and providing formulae and notes for their application. My formulae in many cases are not the original, nor even the classic formulae. I have resorted to easy-to-find ingredients, many of domestic use, raised here to laboratory rank.

Refractive Index

Any time I know the value, I give the Refractive Index (RI) of the proposed mountant. Refractive index is important because it governs the contrast between the detail you are searching for and the background, and also the transparency of the observed sample against the bright field of the microscope. A media with a higher index imparts more transparency. The mounting media must always have an RI higher than the mounted sample. Some aqueous media have an index of about 1.41 (very pure water has an index of 1.33) but Canada balsam has an RI of

1.524, very near that of the glass of slides and cover slips. Naphrax, which is used as a specific mountant for diatoms, whose "frustules" are made of a material similar to glass, has a very high RI of "more than 1.65". There is now even a synthetic media that reaches a really high 1.70+ RI.

Natural media or synthetic ones of RI = 1.5 or more are used routinely in histological mounts of tissues, previously stained and cleared. Most objects (including micro-crustacea, or arthropods) look good when recently mounted in them, but show additional undesirable clearing as time passes by, and many important details, such as setae in cladocera, or copepods or acarii, could become invisible. For these materials the modest RI of the *friendly* aqueous media are actually a better choice.

Selected Mounting Media
We shall review pure water, glycerin, sugars (karo and fructose), gum arabic, gelatin and PVA. I disregard Damar, a very economic alternative to balsam, generally used as a xylene solution, because it is not soluble in any easy to obtain, or safe, solvent. Together, the selected products provide a selection of aqueous media, two of them liquids (antiseptic water and glycerol), with the others—solids, and also one easy to use synthetic resinous medium, NPM (Nail Polish Mountant) that I proposed in a previous article in Micscape Magazine (see references).

Standard Subject
To review the practical solidifying mountant and to provide some comparative images to judge its behaviour, it's best to start with an easy to mount standard subject. And to use a dry one, that needs no fixative, nor any stain to be applied before mounting. I selected fly wings as my, more or less easy to find, standard objects. Of course for some special mountant media I needed and added some alternative test objects.

Equipment
Most of the cited equipment is obvious. But throughout the article I speak about capsules. In professional papers the descriptions would be most probably "watch glasses", "Syracuse glasses" or "cavity blocks". These are useful pieces of equipment. If you have them, or could buy them, don't hesitate, they are the best choice.

But, if you don't have "watch glasses", make a visit to your old relatives. They surely have consumed some medicine tablets sealed in those dimpled plastic sheets. If they take care to not "push up" the tablet, thereby ruining your prospective laboratory equipment, they can provide you with an assortment of sizes of concave plastic recesses (capsules) which are very useful for your laboratory work. Select the largest for the actual purpose. I have concave circular capsules of 10, 12, 15 and 17 mm in diameter, and also some useful ones of 8 x 22 mm.

Here in Durango (Mexico), I have identified one almost ideal large plastic capsule: the caps of one brand of ice-cream cones are plastic cupules of 5.5 cm in diameter. I must make a big sacrifice to acquire one dozen of them, but you understand that science is a priority for me.

Of course don't use them with powerful solvents such as xylene, or acetone. If your materials allows you the use of a white opaque background (indeed it is an advantage in the process of staining) you have recourse to the plates with several concavities, or to the little individual dishes that watercolour artists use to mix their colours.

In addition you may need droppers, tweezers or forceps, fine pointed brushes, wire loops, mounted needles, fine pointed scissors, and a small scalpel. I've searched for but haven't found a substitute for test tubes. There are very useful. If you can, you must buy a dozen or so of 1 cm of internal diameter and a length of no more of 8 to 9 cm. Richard Howey has made a sound review of the laboratory materials an amateur can use more often. Please read his articles. (See the references.)

NOTES AND FORMULARY – LIQUID MEDIA

AW – Antiseptic Water

(Defined here as water containing some diluted fixative, i.e. the water containing the fixed sample.)

This technique has much to do with the usual method of temporary water mounts (wet mounts) you use when studying water samples, microscopic algae and organisms you have just collected in your favourite pond. Except you don't use pond water and live organisms. This is not of course a long lasting mounting medium, but it is useful when you don't want to have your critters drying out, after a long session of microscopic work.

Many times we return from our field trip with some different samples

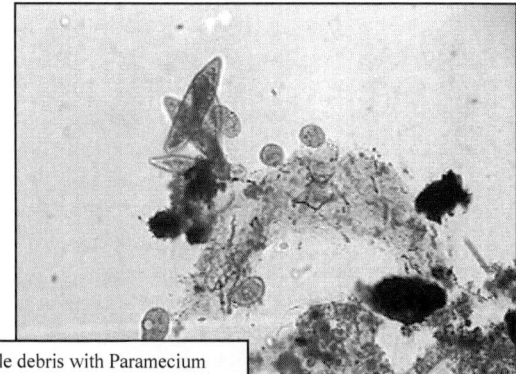

A collection of vegetable debris with Paramecium

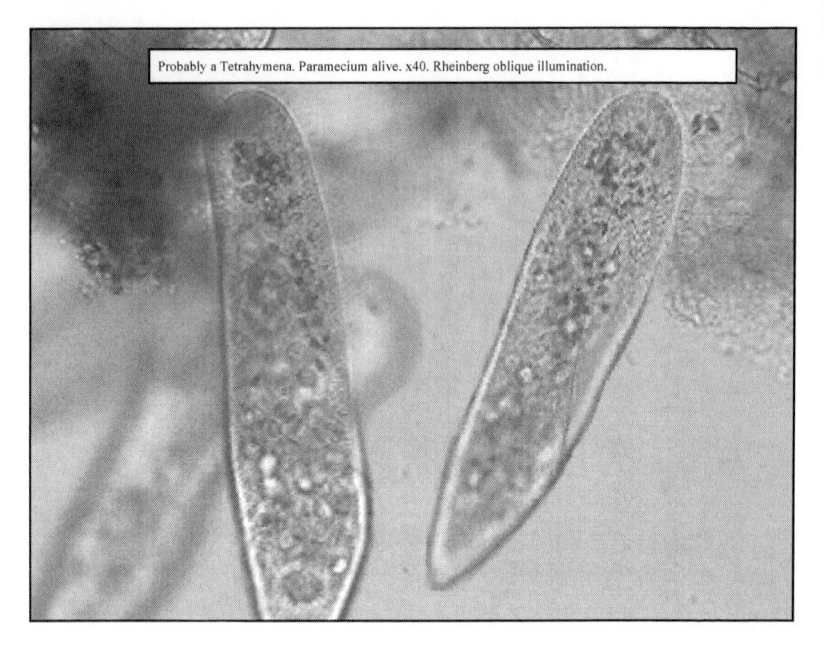

Probably a Tetrahymena. Paramecium alive. x40. Rheinberg oblique illumination.

with one characteristic in common: they are a collection of concentrated planktonic microorganisms, or a handful of detritus, some times very fine, from many possible origins, with probably thousands of interesting….but hidden organisms.

Suppose you have a plankton sample. You mount your drop between slide and coverslip, and start your search. You find many interesting subjects that you want to measure, or to draw or photograph. In a few your sample may be crushed, or lost. You must continuously add water with a fine pointed pipette to the coverslip border…or, better yet, you can seal the water medium to stop evaporation.

Do not absorb the excess water with absorbent paper, this can remove just the critter you are most interested in. Let the preparation evaporate just to the point in which there is no more water outside the coverslip. Now with a fine brush, or the tool I describe afterwards, put one little drop of sealant in each of the four corners. Give the sealant opportunity to set, and continue sealing all the borders.

If you have not "fixed" your critters, with time they will asphyxiate and disintegrate. It would be a good precaution to take two samples. You take home one of them alive and treat it with all the precautions Richard Howey has explained in his recent Micscape article.

You can fix the other using one of the recommended traditional formulae, that, before the new trend for safety were mostly composed of formalin, glutaraldehyde, mercury chloride and other chemicals. They are now reported to be toxic, and not recommended for amateurs.

One useful, effective and safe fixative, that I have designed to fix protozoa, rotifers and the like, has a mild action, and which even preserves the green colour of algal plastids for a while, is:

GALA 20 is less prone to distort delicate organisms such as protozoa. For most of the other micro invertebrate groups use GALA 60.

	GALA 20	GALA 60	60 (professional formula)
lactic acid	1 ml	3 ml	3 ml
vinegar	10 ml	20 ml	1 ml acetic acid
alcohol 96	21 ml	63 ml	60 ml absolute
glycerol	5 ml	5 ml	5 ml
water	63 ml	9 ml	31 ml

Preparing the formula

Don't be deceived by the low concentration of the active substances in the formula. It works. Put 100 ml of water in a suitable flask. Mark the level accurately. Empty the flask. With a 20 ml hypodermic syringe withdraw 1 ml of lactic acid, 5 ml of glycerol, 10 ml of vinegar (that is: 9.5 ml of water and 0.5 ml of acetic acid), and some water. Agitate to mix. Put the mixture in the flask. Syringe out some water, agitate to wash the remnants of the lacto-acetic mixture. Pour into the flask. Repeat one or two additional times. Add the alcohol (21 ml at 96% = 20.2 ml absolute alcohol and 0.8 ml water – more or less). Now replenish the flask with water to the 100 ml mark, and stir until the solution is homogeneous. Alternatively, if you are in the "rich group" of amateur microscopists, use your measuring pipettes and graduated cylinders, to prepare your formula.

Label your flask as "GALA 20 fixative" (because it is composed of Glycerol, Alcohol 20%, Lactic acid and Acetic acid) or "GALA 60" if you have used the higher alcoholic formula.

Fixing the sample

To fix your plankton sample (or any aqueous sample with suspended organisms of the same order of sizes) you must add to it 1/10th of its volume, of this solution. (1 drop to 9 drops, or 1 ml to 9 ml, etc) and agitate well to mix immediately. If you want to fix larger animals (some micro-arthropods, hydracarina, some anesthetized worms, arthropod larvae etc.) put them directly in the concentrated fixative (2, 3 or more volumes of fixative by 1 volume of biomass).

Paramecium fixed and mounted in GALA 60. As I did not use any clearing agent it is somewhat opaque. Nevertheless the exploded trichocysts and the macronucleus are visible.

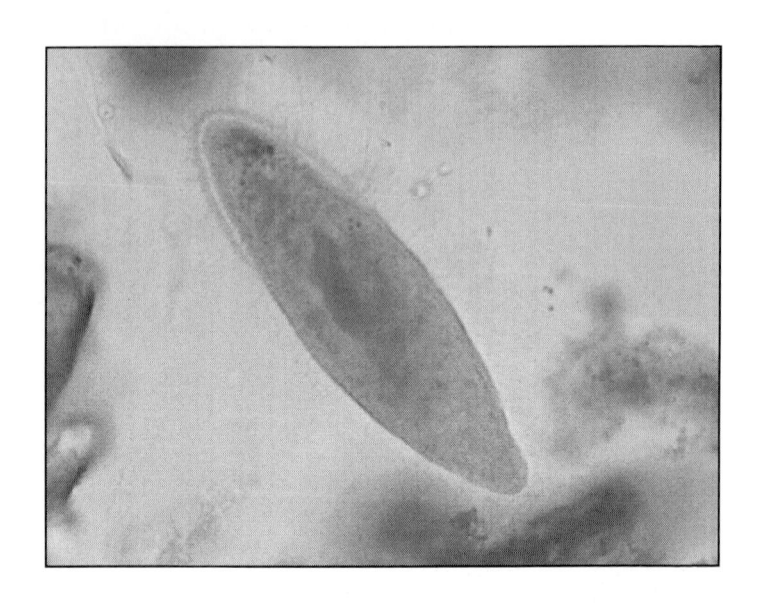

Monostyla lunaris - a rotifer fixed and mounted in GALA 60

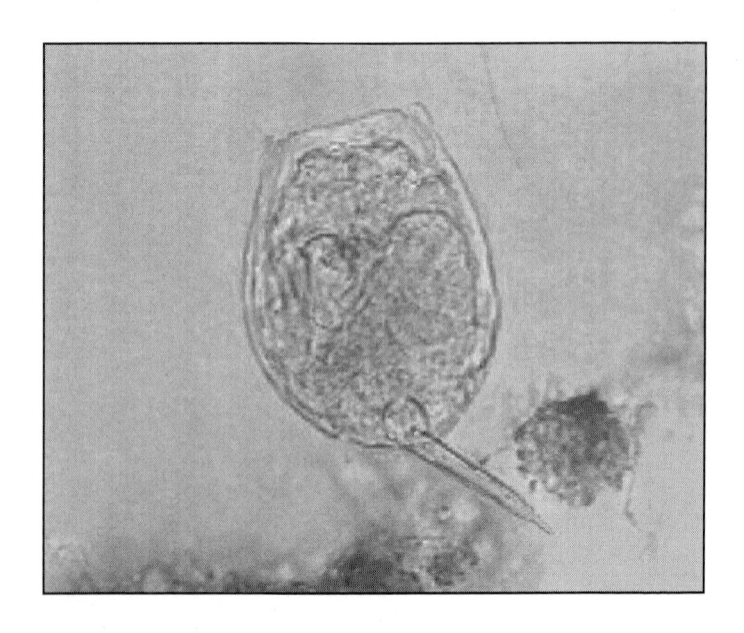

Searching your sample

Of course you can search your materials unstained or apply some colour to better differentiate them from the detritus, or to identify some organelles. One beautiful and useful dye is Rose Bengal, but as for all the good reagents of old times, it is forbidden now. It is toxic. In future articles on staining I hope to review some safe (but inferior) substitutes.

To make your preparation, to aid with the evaporation problem, and to give the subjects a minimal clearing that mimics the live appearance of many micro-invertebrates, mix one drop of fixed sample with a drop of glycerin, or even lactoglycerol (see formula below). Allow to stand for one minute, cover and start your observations. If the materials promise a long working session proceed to seal. You can search by this technique *thecamoebae*, gastrotrichs, rotatoria, nematodes, microalgae and the like. Some specimens, especially some of the protozoa, microoligochaeta and other soft bodied organisms don't support this drastic procedure and become dehydrated and wrinkled. With these materials you'd better proceed to mount in glycerol or lactoglycerol (see formula below) with all the precautions I explain later.

Sealant media

For sealants you can use solid vaseline, paraffin wax or beeswax, Valap (see below), or nail polish. Your choice depends on how long you intend to store your preparation. The selection order is also the order of duration of the different media. A liquid preparation sealed well with nail polish could last some months. The other media need to be applied melted, they remain more or less soft, and can be easily ruined, except Valap, that adheres well to glass, but is also easy to remove if you want to recover your slide, and actually is the other media to be recommended. This is its formula:

Paraffin	1 part in weight
Vaseline	1 part in weight
Lanolin	1 part in weight

Mix

Melt at a low heat and pour in a shallow profile can and cover. Allow to solidify and melt the quantity you need with the following sealing tool.

One sealing tool: you can melt the solid sealing media and use a small fine pointed brush to apply it, but it is very difficult to apply a neat coat to all four sides of your coverslip. An old economic sealing tool can be of help here.

Take a metallic wire, 2.5 mm in diameter, bend the tip at a right angle, to make the bent portion the same length as the side of the cover slip you use most often, and if you wish, make a loop at the other end as an aid to manipulate the tool.

Now, with a spirit lamp, or a cigarette lighter, apply some heat to the sealing wire. Touch the surface of the solid sealing medium. It melts. With the tip of your tool, touch every angle to fix the cover with a little drop of medium. Now apply the heated side of the wire to the medium and to each border. The molten medium spreads evenly along the sides and you have a very good seal all around. It takes only a few trials to become an expert. Use this tool with Vaseline, the waxes and Valap.

Comments: *These simple, temporary mounting methods, are probably the ones which you would use most often, hopefully with much success. They are useful with hydracarina, thecamoebae, nematoda, loricate rotifers, most of the Gastrotricha, Chlorophycea (which retains its green colour for a long while), euglenoidina (many of them showing his flagellae), desmids and Cyanophycea.*

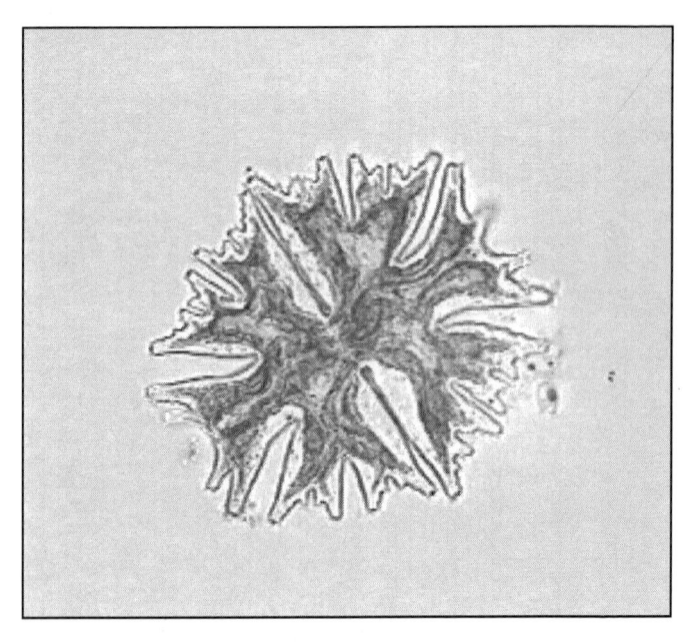

A desmid retaining the colour of its chloroplasts after two weeks fixed and mounted in GALA 60.

The entomostracans (cladocera and copepods) have calcium in their exoskeleton that could be dissolved by this highly acidic fixative. So you'd better fix and preserve them with 70% alcohol. If you fixed them with GALA, because they were mixed with other critters, select them after a few hours, and store in 70% alcohol.

Ciliates can be well preserved, showing their nucleii (macronucleii at least), and cilia. Those that have trichocysts, as paramecium do, show them exploded most of the time.

Only for comparatively long lasting collections, morphological detailed studies, and demonstration purposes, or when your materials need clearing, as in many arthropods for example, you must resort to the more dense, permanent …and more difficult to use mounting media.

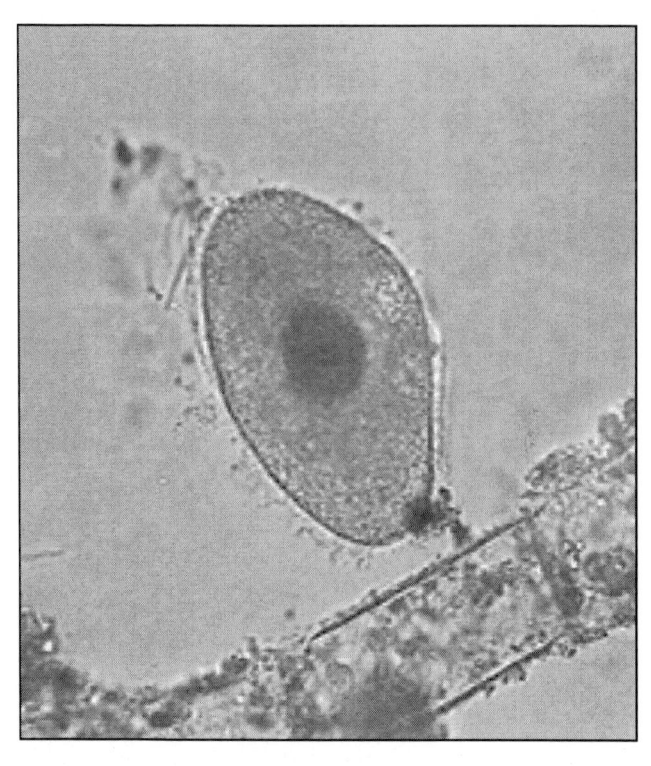

(Above) Probably a Tetrahymena.

(Below: next page) A couple of conjugating Paramecium. Both fixed and mounted in GALA 60. The pictures increase the general opacity and the contrast of the nuclei. They are less contrasty, but easily visible in the mounted materials, without using dyes or clearing agents. A note on mounting samples in aqueous media: Quite frequently, water evaporation applies a high pressure to the subjects with the risk of delicate subjects being crushed. To avoid this you must resort to the inclusion of some supports for the coverslip. Those supports must be thin enough to permit a high resolution objective to be used, but thick enough to give room for the observed subjects.

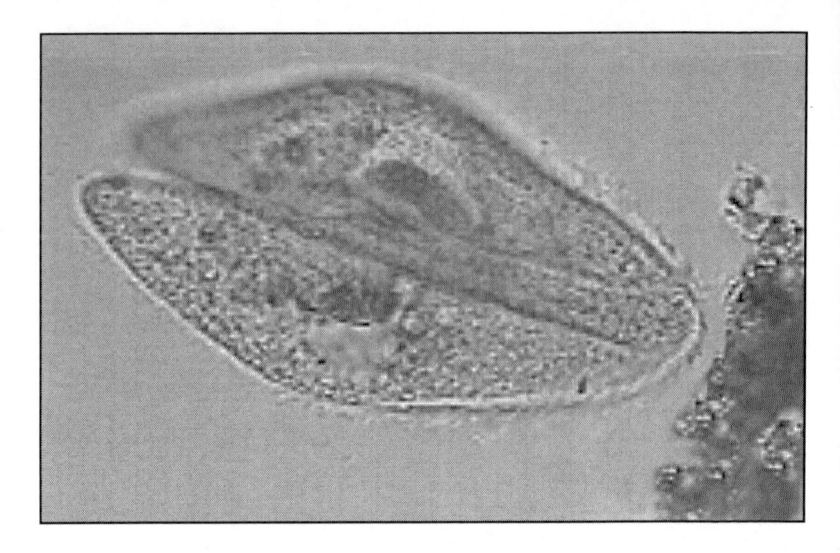

Use at will pieces of thin coverslip, paper, plastics, hairs or textile fibres. Many thin adhesive tapes can be cut into thin strips and neatly adhered beforehand to the slide.

Think before you have resort to these methods, because you can't revert the situation. If you think that it is possible that you may change your mind and want to slightly squeeze an organism, it is best that you use, instead of any of the fixed height materials, tiny balls of very soft wax, or drops of solid Vaseline. These allow you to put a more or less controlled pressure on the coverslip, to stop the displacement of an organism or to reveal more clearly its internal organization.

Labelling.
Even with these not so permanent preparations you must label your slides. At least write the name of the mounted subjects, the source of the materials, any treatment applied, the mounting media, and the date. Add your name if you want.

Even if the slides are to last for only some weeks or a couple of months, when you make a revision you may have forgotten those details, and be sorry you did not label them. You can make the labels on your computer. But you must write the information with a good indelible black drawing ink...or perhaps you could finish with an illegible label caused by accidentally wetting it.

MOUNTING IN PURE LIQUID GLYCEROL

PG – Pure Glycerol

Glycerol, also known as glycerin, is a common product, cheap, and easily acquired in any drugstore. It's a very hygroscopic alcohol, with a weak syrupy consistency and when anhydrous has an RI = 1.46. Buy well sealed small bottles of glycerol and open it only for the time needed to take out the drops you use.

It is most probable that glycerol was used, as a temporary mounting medium, from the very beginning of microscopic techniques. It is most easy to use. Put a drop of glycerol on a slide. Include the test object (our fly wing), removing it from water, or alcohol, cover…and go to the microscope.

Many microscopists seduced by this simplicity, the very high compatibility of glycerol with many solvents, and the added fact that it is a mild clearing agent, that imparts a fair transparency to the small biological materials it impregnated, tried to make it the mounting media for their permanent preparations.

The straightforward method is to seal the glycerol mount. It is not at all easy, but it is possible.

When you want to turn your glycerin mounted slide into a more or less permanent preparation, put the slide on the table over one or two sheets of paper. Cover it with one sheet of absorbent paper (thick kitchen towels work, or even toilet paper) and slide the edge of your hand from end to end of the preparation, taking care not to exercise an excessive pressure. This squeezes the excess of glycerol from under the coverslip into the towel.

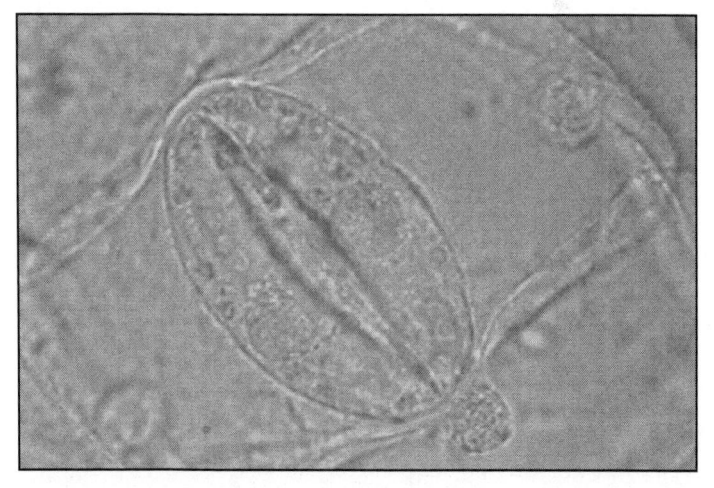

Stoma on the epidermis of the underside of a leaf of *Tradescantia*

virginica. Fixed in AFA (alcohol, formalin, acetic acid). Mounted in glycerin. No stains applied.

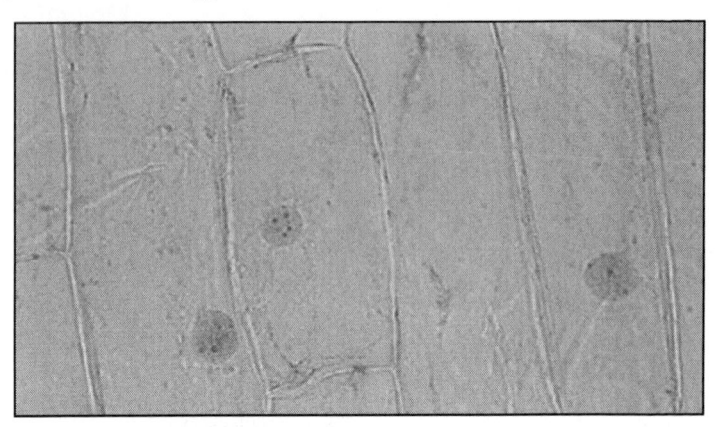

The well known epidermis of an onion. Fixed in AFA. The staining method will be explained in a future article. (chapter in this book). Mounted in glycerin.

Uncover the slide very carefully, wrap your finger in some thin non-absorbent plastic sheet, support one side of the coverslip, and with extreme care wipe with a dry cloth all the oily remnants you can. Now you need to seal the coverslip.

First of all put one drop of nail polish on each of the four corners, and allow to dry completely (not less than half an hour). Now support your coverslip, moisten a cloth with a little alcohol and wipe very carefully all the contours of the coverslip. Any rough movement and you'll ruin all your work. Use a dry cloth to finish. Both the slide and cover slip must be dry, with no glycerol present at all.

Now seal all four sides with a nail polish layer that overlaps more or less 1.5 mm of the cover, and on the slide. Or use your sealing tool and Valap. Particularly take care to also cover the four corners you fixed earlier. Allow to dry completely. Wipe with alcohol, and apply another, very carefully applied, sealant layer of NP. Normally Valap won't need a second layer.

Make a check next day. Glycerol is highly hygroscopic and absorbs water greedily. If there is a minimal opening in your sealant there'll be a glycerol-water mixture flowing out. Dry, wash with alcohol, seal the opening. Make a daily check for one or two weeks. When you are satisfied there are no more leaks, finish the seal with automotive or hobby paints.

As the mounting media is a fluid, you must file these preparations in flat horizontal trays, otherwise the subjects can be displaced. Carefully label them of course.

As to the permanency of these glycerol preparations, it's worth pointing out that J.G. Baer, a French helminthologist, reviewed in 1931 the taxonomic characters of Temnocephala mexicana, Vayssiere, 1899 from the type slide mounted in glycerol, and filed in the collections of the Museum d'Histoire Naturelle, at Paris, France for 31 years. (T. mexicana is a platyhelminth that lives on some crabs.)

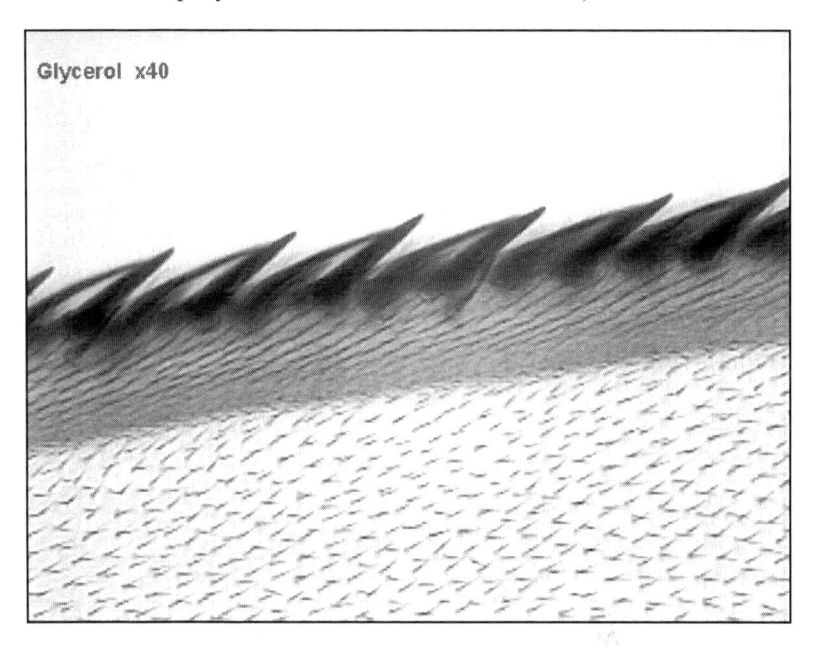

Glycerol x40

(Above). This is the first picture of the test object. Fly wing collected and preserved in alcohol 70%, and mounted in glycerin. (Below) Head of the larvae of a Sarcophaga fly. Fixed in 70% alcohol, it was gradually passed through increasing concentrations of glycerin and mounted in 100% glycerin.

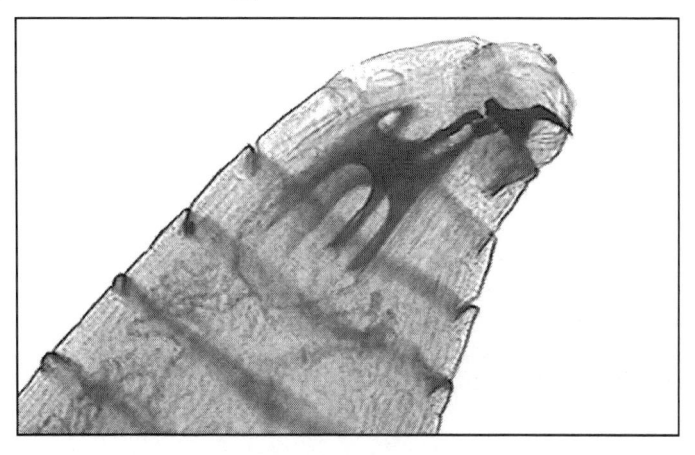

Difficult materials. As I said, glycerol is hygroscopic. It acquires water from all substances that contain free water. When a soft biological object is surrounded with glycerol, water will pass from the object to the glycerol, and is replaced by it. The problem is that glycerol diffuses to the biological materials at a slower rate than when water goes out. The results of this imbalance is the biological materials are shrunk and distorted. To make a useful mount of this kind of subject we must resort to two useful tricks.

1) In a series of capsules put a series of increasing concentrations of glycerin, starting with the 10% side of the scale (20, 50, 75 could be the other steps). Put your materials in the first one and leave enough time. (Of course you must guess, and try some times, until you find the correct one for your actual material. Twenty or thirty minutes are wise initial guesses.) Using a suitable spatula or a little brush or a wire loop, or even an eyedropper, transfer the materials with care from one capsule to the next, leaving them for the same time in any one. Perhaps after more or less two hours you are done. Transfer the subject in the drop onto the slide, as was explained before.

2) Often the first alternative is good enough. but Seinhorst (1959) working with difficult nematodes, designed a technique to gradually replace the fluids with glycerol without disrupting the organ structure of the worms. By the same nature the method is excellent for any delicate material.

The materials, first collected in water, were then transferred to a capsule with a 1% solution of glycerol in 20% alcohol (more or less).

    alcohol 96...................21 ml
    water........................78 ml
    glycerol......................1 ml

Now take a wide mouthed flask, with a screw cap, not very tall. One of those short and wide containers that hold creams for the skin or hair fixatives, are good. Put in the bottom a cap from another smaller diameter flask, as a platform to support the capsule. Pour in 96% ethanol almost to the height of the platform, and put the capsule with your materials on the platform. Spread some solid Vaseline on the rim of the flask. Screw on the cap of the "interchange chamber" tightly and set aside for 12 to 24 hours in a warm place. It's better if you apply some heat, say 35-40°C. In the interim all water in the capsule was to be replaced by alcohol.

Now fill the capsule with a solution of 5% glycerol in 96% ethanol, and put it in a partly closed container. In 3 or 4 hours at 40° almost all ethanol has evaporated, and your subjects must be in almost pure glycerol. Normally you can now proceed to mount in pure glycerol.

3) One special case is the mounting of mixed microscopic microinvertebrates, which because of their minute size cannot be manipulated individually, or in sorted uniform groups, and must be mounted in some liquid media. (The samples we spoke about in the "antiseptic water" section). The following, although not easy to apply, is a useful protocol.

a. Fix and preserve the materials in a suitable fixative (GALA is a good alternative). Allow at least 6 hours for a good fixation.

b. Have a look at the sample to see if they are the critters you are searching for. If there is a need for some staining (await a future article) this is the moment to apply it.

c. Take a sample (say 19 drops). Add 1 drops of glycerin mixing very well after the addition. You now have a solution with 5% glycerin. Set aside for 15 minutes.

d. Add a new drop of glycerin, mixing well. The concentration is now near 10%. With a syringe and a fine hypodermic needle remove half of the supernatant liquid. If you work with care you now have 10 drops with more or less 9 drops of sample water and 1 drop of glycerin. (10% glyc.) Add 2 drops of glycerin and mix well. This amounts to 12 drops (9 + 3) of 33% glycerin (more or less). Set aside for 15 min.

e. As the density of the liquid increases, so the sedimentation time is longer. Give enough time to have almost all of your sample in the test tube or concave capsule bottom.

f. Now take one clean slide and put one drop of the concentrated sediment in the centre. Mix well with another drop of pure glycerin or lactoglycerol (see below), and cover. You must make a fair estimate of the volumes to have a very thin preparation but full to the margins of the cover.

g. Seal if you wish. The method is not quick, nor easy, but gives you pretty good preparations of many difficult to preserve microinvertebrates

NOTES: Make your preparations as thin as you can. Resolution is impaired in thick liquid preparations, and also it affects the quality of the sealing. Always allow enough time for the sealant to set completely. Never use the 100x OI objective (if you have one), with unsealed preparations.

LG - Lactoglycerol

I should point out that glycerol has a mild clearing power, enhancing the transparency of fixed cytoplasms and other small opaque subjects. The combination of glycerol with lactic acid (a more powerful clearing agent) produces a medium that is very useful to allow a better observation of opaque materials and organisms, like worms, plant materials and especially small arthropods.

Lactoglycerol is normally a 50/50 mixture of both reagents, but you can freely use other percentages according to the clearing action required by your subjects. It is most often used as a temporary mountant, but with suitable precautions, a satisfactory preparation can be turned into a permanent one. The manipulations are much the same as with just glycerol, but sealing is more difficult.

Lactoglycerol is more dense and syrupy and is more difficult to clean the outside of the coverslip. You must anchor the four corners, waiting until the sealant is very well set. With an absorbent paper, wipe all the excess liquid you can. With a soft absorbent tissue wetted with ethyl or methyl alcohol or even acetone, clean the slide with care. Always try to have the thinnest preparation that you can. When you are satisfied of the cleanliness you have attained, seal with Valap. Inspect after a few days to be sure you have no break in the seal. Cover Valap with a layer of nail polish and finish with another of an automotive paint. Label your slide.

Pantin's solvents density gradient.- If you like to mount a micro-arthropod (a mosquito larva as an example) you must clear the organism if you want to see the details of its structure. One of the methods is to dissolve the soft parts (all the internal organs) with hypochlorite or potassium hydroxide. Both are potentially harmful.

A safer method, usually enough for our purposes is to use the lactoglycerol mixture as a clearing agent. If you submerse the subject directly in lactoglycerol you know the result: a wrinkled and awful organism, mostly unusable. You can resort to gradual changes of higher concentrations of lactoglycerol.

But Pantin at the end of the 1930's proposed a very useful and elegant method that provides organisms beautifully cleared with only one (long) easy step.

Take a test tube or similar, put 2 or 3 ml of lactoglycerol at the bottom. With all your care, with a long pipette, glide 2 or 3 ml of glycerol over the first layer. You can disturb it a little but only at the very interface to start a process of mixing. Now in the same way, make an upper layer of 96% ethanol.

With your tweezers, or a glass bar, or an entomological mounted needle, pick the subject from the alcohol in which you have it stored, and drop it on the alcohol upper layer. The organism drops to the interface of alcohol-glycerol and stays floating there for a long while.

If it is your first time you'll certainly want to watch the process, but better you go to take a nap or make other observations you were planning.

The larvae or other organism you leave bathing in the alcohol slowly mixes with the layer of glycerol, interchanging the alcohol, and after a time it sinks to the interface of glycerol and lactoglycerol where it repeats the process.

At the end you could recover the organism from the bottom of your tube. Perfectly cleared and with much less distortion.... and requiring less effort from you.

Experiment, and become addicted to the method.

The organism you have cleared can be mounted in lactoglycerol or in some of the more solid mountants I describe in the next chapter.

Four pictures show the clearing power of lactoglycerol. The mosquito larva (above) was fixed in 70% alcohol, and cleared using the

solvent density gradient of Pantin. The solution layers used were alcohol 70% - glycerin - lactoglycerol. The sunken larva (top—next page) was mounted in pure lactoglycerol. The third picture shows the nucleus of the cells inside the caudal paddles. The fourth one shows two muscular cells at the end of the body. The third one shows the nucleus. The fourth one shows that these are striated muscle cells. Focus was a compromise to show both details in one picture

Another method for using glycerol as a mountant.- As you see above the biggest problems with glycerol and lactoglycerol arose from

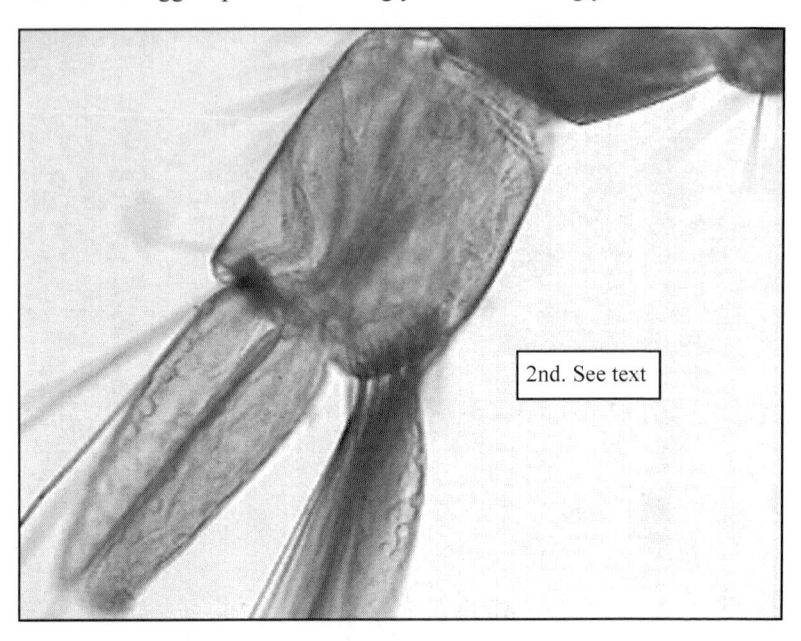

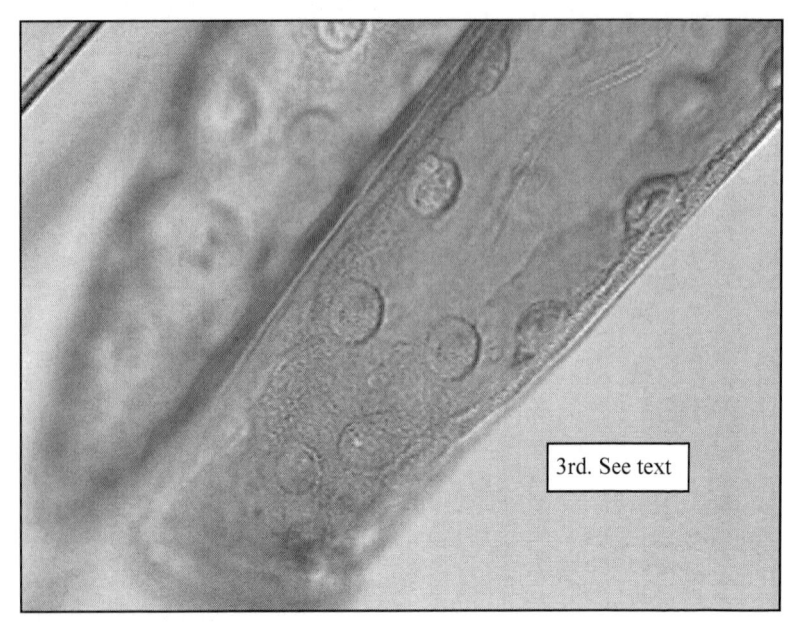

2nd. See text

3rd. See text

3rd. See text

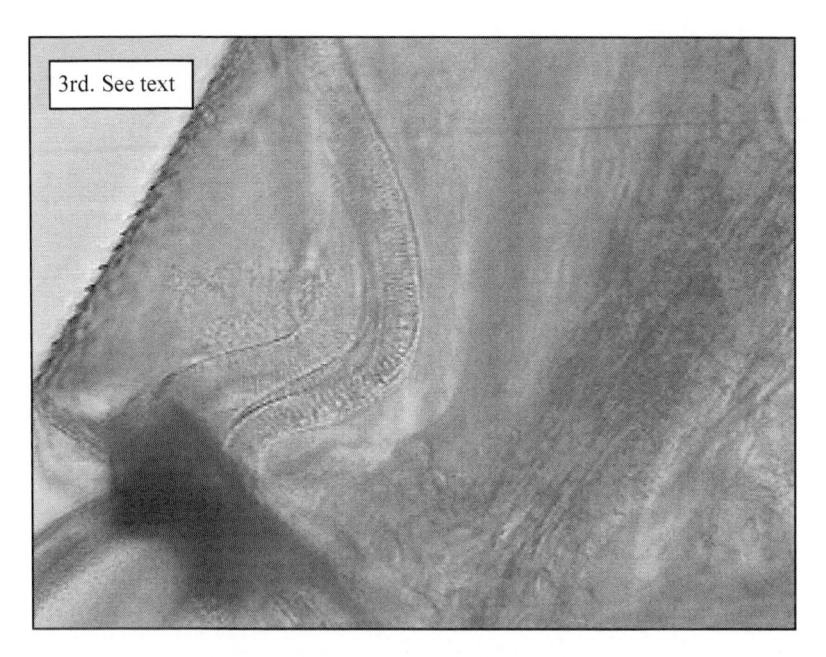

their nature as being viscous and hygroscopic liquids. Microscopists solve this by converting glycerol to a solid. The commonest solid form is glycerin jelly. I have two other useful formulae fructogel and glycogel.

All of them are explained in the chapter three of this book. In the next (second chapter) I will explore the use of gums and sugars alone or in several combinations.

References

Walter Dioni. About microscopy and the chemistry of nail polish.
Richard Howey. Equipping a laboratory (parts I to IV).
Richard Howey. Culturing and collecting microorganisms safely.

Note: These reference refer to articles on the Microscopy-uk's web site and its daughter magazine: Micscape. Richard Howey is another long time author whose work is much published both in Micscape Magazine and other works.

Web Address is:
www.microscopy-uk.org.uk

And look in the search database on the Micscape page.

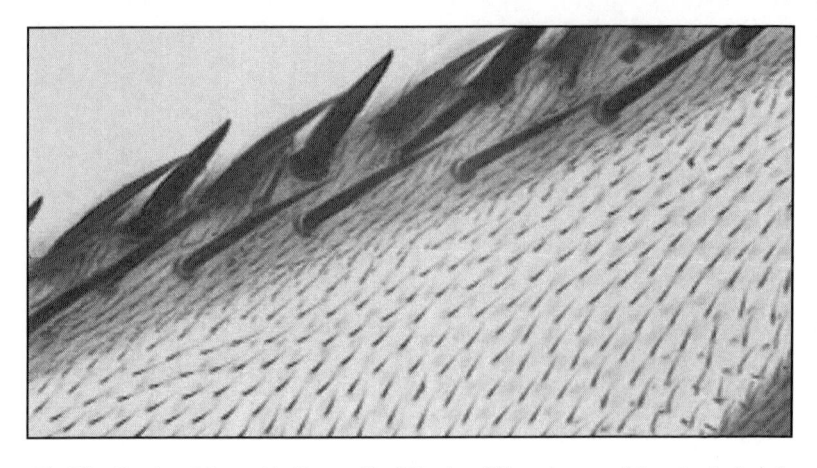

Fly Wing Border - Mounted in Karo - 40x Objective. This and many of the other included images have been amalgamated with CombineZ (Focus stacking software). All the original pictures have been captured at a resolution of 640 x 480 and cropped or reduced as needed to include them in the article.

MOUNTING MICROSCOPIC SUBJECTS.
Chapter 2 - Solidifying Media

Of the three more common sugars (sucrose, fructose, and glucose) the first, which is a disaccharide, was used at the end of the XIX century in von Apáthy's mounting formula, based on gum arabic (also known as gum acacia, a mounting medium used in microscopy since 1832). But as for the content of sucrose, it is said to crystallize ruining the preparations, and also, in its original formulation, fades most of the usual dyes, so it was reserved only for one special application: as a mountant for histological sections of tissues with lipids stained in Sudan III. The original formula is not a most promising one for the amateur's laboratory.

Lillie substitutes fructose for sucrose in von Apáthy's formula, to inhibit to some extent the crystallizing behaviour, and includes potassium acetate to enhance retention of other dyes. The gum arabic-sucrose formulas have a low refractive index, about 1.42, but Lillie's fructose based medium is said to reach 1.46.

A glucose syrup at 98% concentration was included in two other gum arabic formulas (Berlese and Doeschtman).

The gum arabic media were popular with arthropod experts. Apart from the just cited authors, Farrant, Dahl, de Fauré, Hoyer, Morrison, Olsen and others also contributed more or less successful formulae. Of these the Berlese, de Fauré, Doeschtman, and one of Hoyer's formulations are banned from the amateur and from many of the professional microscopists laboratories because they include chloral hydrate, a toxic substance now forbidden.

Chloral hydrate is a (toxic) highly refractive reagent that clears the tissues, making them transparent, and is specially useful to reveal the internal structure of dissected parts of arthropods, micro-arthropods, and microscopic worms. It is very difficult, even impossible, to replace chloral hydrate. Once a useful substitute, or better a complement, was phenol (carbolic acid) but it is also forbidden now. Chloralphenol, the powerful clearing fluid derived from the above reagents, had a powerful RI of 1.54.

Discarding chloral hydrate and phenol, and substituting fructose for sucrose some gum arabic formulas can be useful for the amateur, but there is an additional problem. Gum arabic use to be supplied in two solid forms: as a powder or as lumps. Most microscopists agree that the powder is not useful (it's almost impossible to dissolve it adequately, and mostly extended with cheap products). You must try the suppliers to candy makers to have a chance to get some in solid lumps of the required quality.

As an amateur I circumvented the problem by using in my formula a commercial solution of gum arabic, readily available over the counter for watercolour painters. But I don't know if this is a solution that everybody can apply, in every location. You must search the suppliers of products for the artists, where some gum arabic solutions of a good syrupy consistency are sold. Some of them may be too thin, and possibly you would need to concentrate it by leaving it open in a warm place to evaporate off some solvent water.

Fructose without other additives, was proposed by Lillie as a concentrated syrupy solution useful as an aqueous mountant. A solution of 75g fructose in 50 ml of water is rated by Lillie to have a good refractive index of 1.476.

Larry Legg gives step by step instructions to make a very useful pure fructose mountant *(See further on in this chapter!)*. Concentrated corn glucose, when it started to be sold, and now a high-fructose corn syrup, Karo™ is a commercial syrup that was proposed a long time ago (surely before 1937) as a mountant for mycological, physiological, and botanical materials. It is an easy to find mountant. Buy it in any supermarket in the "Light" or "Clear" form. I consulted the manufacturers about some of the characteristics of the product and they kindly answered my e-mail. Karo Light is sold as a solution with 76% total carbohydrates. At 20°C it has an RI of 1.484 to 1.486, very high for an aqueous mountant.

Of all the media discussed here, Karo offers the simplest solution.
Dave Walker informed me that Karo may not be common (at least with this name) in the UK, and possibly elsewhere. In this case Larry Legg's formulation is surely the best alternative, or you can search your shops

to select a commercial syrup made with fructose, (from corn), and not sucrose, (from sugar cane). Lyle's Golden Syrup, a popular trademark in England is in fact a syrup of sucrose (common sugar from sugar cane), glucose and fructose. (From the maker's web site a typical mixture is ca. 30% sucrose, 50% invert (fructose/glucose)). I think that this gives the opportunity to put it on trial to assess its behaviour. If it doesn't crystallize in a reasonable term of several months, it could be selected as an amateur mounting media. Most of the commercial syrups are of a golden or brown colour to imitate honey. If the almost colourless "Light" presentation is not obtainable, the golden ones can be useful. Remember that Canada balsam is more or less yellowish.

I intend to review here the prospective solidifying mounting media that, like glycerol itself, can be used straight out of the container, without needing to be mixed with other ingredients, but solidifies at least at the margins of the coverslip, making it much easier to make the reparations.

A little word of warning. When you mount in AW (antiseptic water, you remember) or PG (pure glycerol) you can scan your preparation with your microscope immediately. When you review the slide some weeks or months after, the most probable event is that you see your subjects almost as you saw them the first time. The media we'll study in the future have a different behaviour.

Almost all of the solidifying media need some time to penetrate completely the subjects you mount, at least if they have a certain volume. So the appearance some minutes after mounting is different (sometimes very different) than after some hours, days, or weeks have elapsed. Be aware of this and take advantage of this behaviour.

Examine your recently prepared slides many times. At first every few hours, and after this every few days, until you can see a stabilized slide. Record what you see, draw the new details you can discover, make your documentary pictures. All this adds to improve your knowledge of the subject (and of the mountant). If you don't do that, you can be missing important information. This is more true of the media that have a clearing agent included, and are discussed in the next part of this series.

Using either Fructose or Karo™

Techniques to use both media are the same. The only difference is that you need to prepare your own fructose solution, and when you do it, you really know what is in the flask. Follow Larry's instructions (below) using crystalline fructose. Karo (or similar) is not so pure but is useful, and is ready to use.

FMM.- Fructose mounting media and KARO .- You can use Karo directly from the bottle, you can dilute it, or concentrate it if you want to mount some large pieces. You can mix it easily with other ingredients such as glycerin. It is optimal with inanimate subjects, or with chitinized parts. Phycologists, mycologists and some botanists use pure Karo as a mountant. It deserves more attention from zoologists.

Note: A phycologist is a person who studies algae. I give a detailed report below about the use of this medium, but it is not because it is the best, it is only the first. And almost all the techniques explained are useful with the remaining solidifying media.

Larry Legg's advice on Using FRUCTOSE SUGAR as a mountant
This is our magic ingredient which helps to make slide-making more accessible to younger people or, in fact, even to long-standing Microscopists. We are going to make up a stock solution ready to use for making loads of slides. Remember though that because we are going to make slides using a SAFE method, we are going to skip a couple of important processes which would require the use of chemicals not quite as safe to use for youngsters. This will mean that some specimens will not be perfectly mounted for absolute clear 'viewing'- but we should end up with slides acceptable to the beginner, who may then wish to move on and adopt the processes omitted here to refine his or her future slides!

Pour some Fructose sugar into a clean dry glass jar. Ideally, you want to fill the jar two thirds up to the top. Pack the sugar down so it is firmly taking up the space in the lower part of the jar. Take a pencil or felt-tipped pen and mark the level of the sugar in the jar. You do this by drawing a small dash on the outside of the jar exactly where the top of the sugar reaches to.

Gently pour some de-ionised water into the jar. The sugar is very soluble and will start to dissolve immediately causing the level to drop. You must keep adding water to maintain the level to your original mark. When most of the sugar seems to have dissolved, you will notice the liquid is not yet clear because it retains crystals of Fructose which still have not dissolved. Put the cap back on the jar and leave it in a safe warm place overnight. This will allow time for all the sugar to dissolve and you should end up with a nice clear liquid.

Important extra step! You can use the Fructose in this strength (as mixed above) but I have found it easier to mount the specimens I've done so far in a slightly weaker solution. Some things tend to 'float' on top of the thicker solution and if you end up getting air bubbles in the drop you use on the slide, they are more difficult to get out. I recommend you: add more water to the solution at this stage to raise the level to half way between your mark and the top of the jar! Put the cap back on firmly and turn the jar upside down and then right way up a few times to gently merge the water with the syrupy solution. Leave for a few hours before using it to ensure the sugar stock 'thins' out throughout the jar. *Larry*

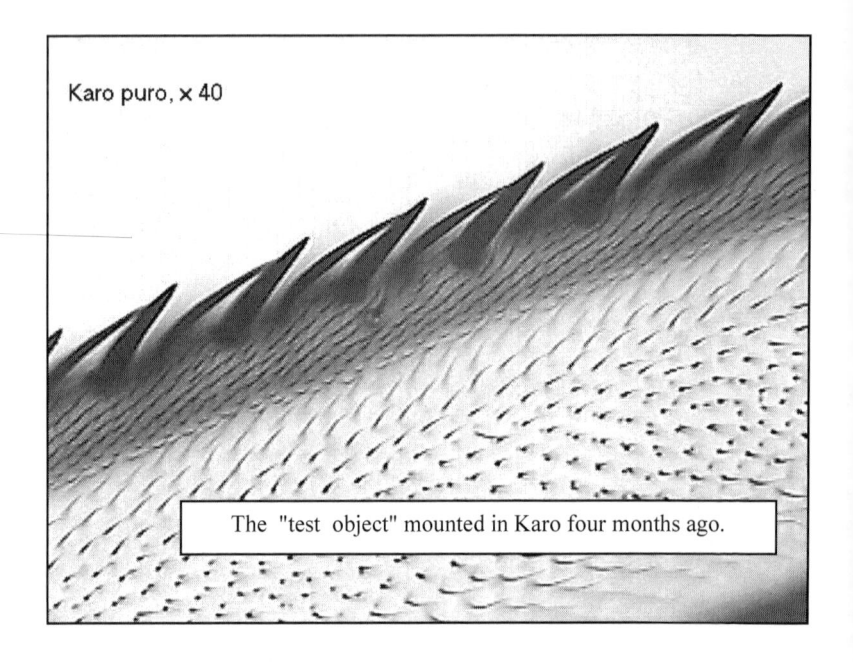

Karo puro, x 40

The "test object" mounted in Karo four months ago.

Pre-treatment
For dry objects the pre-treatment is a way to moisturize them, thus making it easy to mount in Karo. Pour alcohol 70% (common rubbing alcohol) in a little capsule, and water in another. Mix with care so as not to include air bubbles, two levelled teaspoons of Karo with eight teaspoons of water. Pour a little of this 20% Karo solution into a third capsule.

Now with your tweezers, or other appropriate tools, take the object through the 3 capsules leaving it for 30 seconds in each of the first two and one minute in the latter one.

If your subjects are fixed, and are in a liquid, the steps depend on which liquid it is. If it's alcoholic, make a first washing in the same degree of alcohol of the fixative. Generally you part from a fixative with 70% alcohol; so, wash in 70% alcohol, and continue through water and 20% Karo to the pure Karo mountant. If it is an aqueous mountant as GALA or similar you skip the alcohol and pass directly to water, and 20% Karo.

Mounting

Put a little drop of 50% Karo or of pure Karo on a slide. Use a glass rod of 2 or 3 mm of diameter. If you use a coverslip of 22 x 22 mm, the drop you dispense must have a diameter of 2.5 or 3 mm. Use the concentration best suited to the size and kind of object you are mounting. Normally 50% or 75% are the best concentrations. For the fly wings I selected pure Karo. Put another somewhat smaller drop on the coverslip.

With the forceps, recover the subject from the last capsule and slide it into the mountant drop. To help you, use a mounted needle if you wish, and layout the object in the drop. Take the coverslip and invert it with a rapid movement to leave the drop hanging from the underside. Now put it in touch with both drops, and let the coverslip go down under its own weight, with the mountant flowing homogeneously up to the borders. For the object to lay down flat, in the minimum amount of mountant, and to assure a better resolution when you see it under your microscope, put on the coverslip a weight of more or less 10-15 grams. (Take some 2.5 cm long, flat-headed screws, of 3 or 4 mm in diameter, and screw-on as many locknuts as you need, 3 or 4 are just enough.)

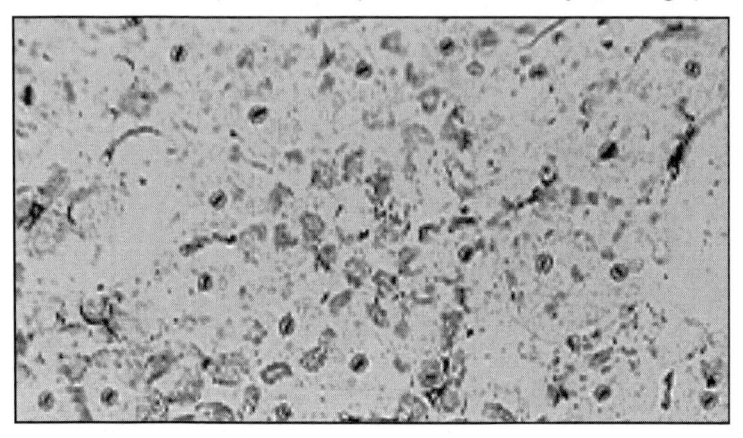

Epithelium from the underside of a leaf. (Objectives x10, x40, x40, x 100).
Fixed for 12 hours in lactocupric fixative.

The second picture [2] (below/next page) is a montage in glycerin for comparison. The other two specimens shown here were mounted in Karo, through 3 increasing concentration steps. The preparations were 14 days old. Note the good cooler conservation of the chloroplasts and cytological detail, including the cell nucleii.

This procedure is not as easy as it seems if you use concentrated mountant. Fructose, and almost all the other thick media, are dense enough to oppose the sinking of your subjects, and even of the coverslip. The first drop must be small and the object must be sunk with your tools to the bottom, to prevent unexpected and erratic displacements of the objects when the coverslip is applied.

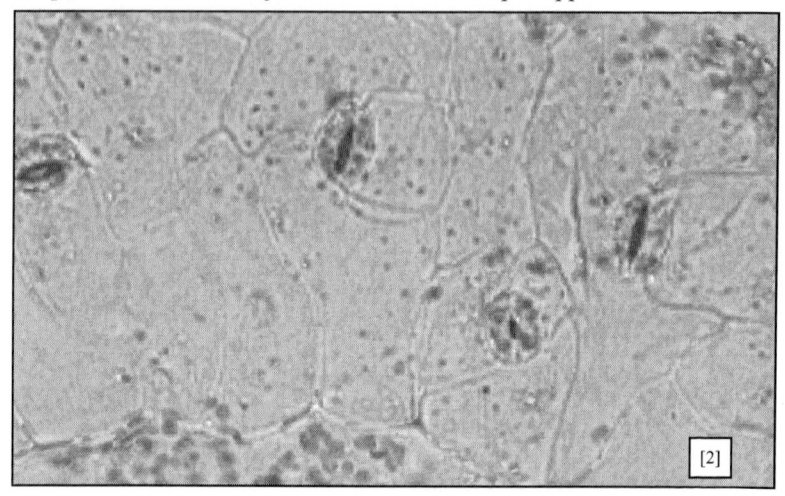

Montage in glycerin (above).

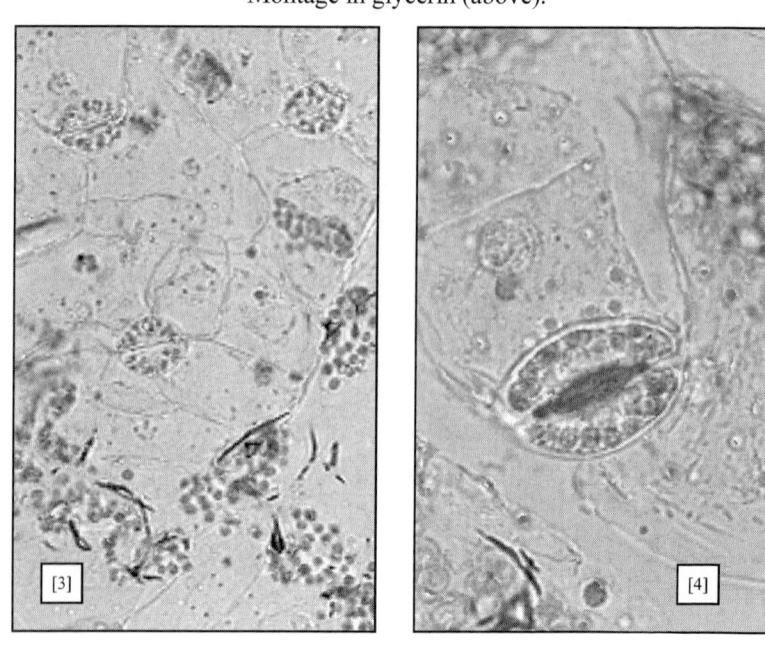

These two mounted in Karo

Before attempting to mount a good subject make several trials to learn the best concentration of the solution, the appropriate diameter of the dispensing rod, the ideal size of both (slide and coverslip) drops, the amount of hand pressure you can apply, and the delays you must allow between each operation you do. Allow enough time (half an hour at least) for the preparation to set before applying weight over the coverslip.

Drying in the air.- Leave the slides on a flat and horizontal table to be air dried, under a cover to protect them from dirt. Depending on ambient relative humidity the borders can be dry in 2 or 3 hours, but sometimes, in some climates, you could need one or two days. To set well could take more than a week. Don't be impatient. Of course, with due care, you can view your slides under your dry objectives, but do not attempt immersion before a week.

Drying in the oven

To speed the process, preparations are dried by professionals in a laboratory oven at 40°C. Forty degrees (and more) can easily be attained, in an amateur's style, by putting the slides over a desk lamp with a 40 or 60W bulb. Turn it on, put a thermometer between the preparations, and adjust the height of the bulb to obtain a reading of 40°C. The process is more efficient and economic if you surround this setup with a wall of cardboard or other insulator.

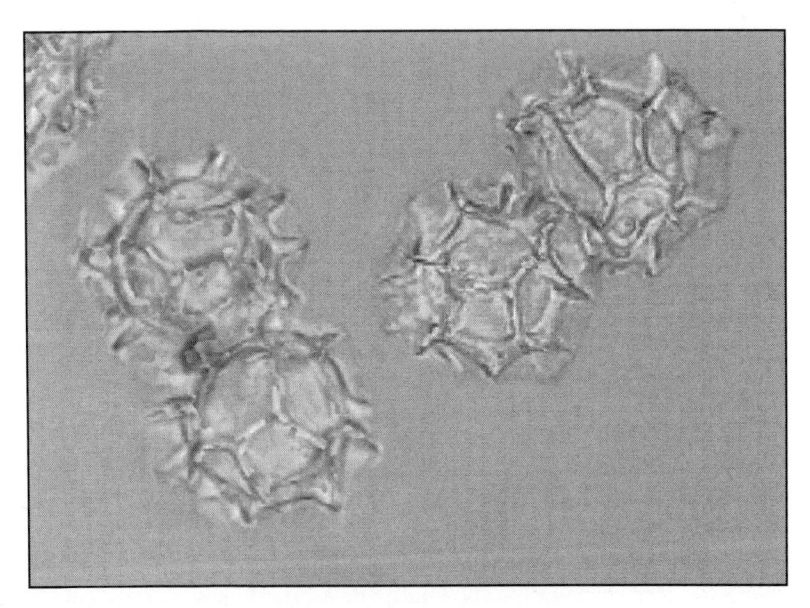

Pollen grains, objective x100.

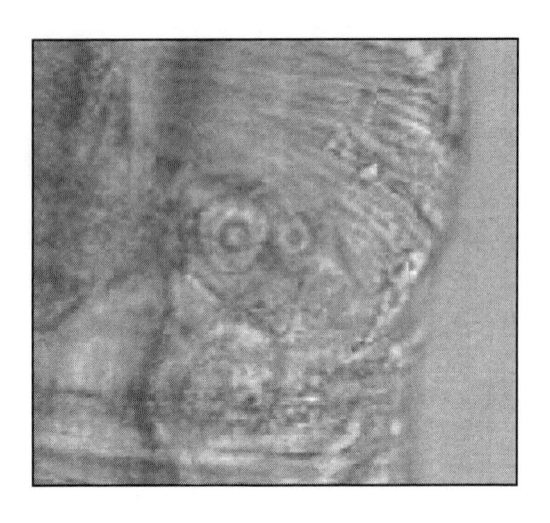

Above. Cells of the ovary seen through the exoskeleton of
a diaptomid copepod.

Microwave oven
Amateur microscopists who possess a kitchen microwave oven can find it profitable. In my 700W oven, 30 seconds of continuous radiation at FULL (100%) setting provides a fully charred slide. But a 10 - 15 seconds exposure dries it beautifully. Be aware of a curious fact. When you heat your slide, Karo becomes fluid, and if it is in a thick layer it can move under the coverslip displacing your subject. Make your preparations thin and give your slide many short periods of heat instead of a long one.

You must experiment with your microwave. Be aware also of the fact that many domestic microwave ovens have a non-uniform distribution of radiation. Try several positions of the slide on the turntable. Use a scalpel, a Gillette blade or an X-act to shave as much off as is possible the dry and sticky Karo that exudes around the coverslip. Give another 10 seconds treatment. With a piece of cloth moisturized with tepid water and/or alcohol clean the cover margins.

Sealing the dry preparations

Many authors state that sealing is not needed at all. They must live in very dry climates. Others use nail polish or automotive paints to seal the borders. With a flat brush with a squared point take a drop of the sealer. Apply the brush, to the margins of the coverslip and paint a line that overlaps 1.5 mm of the cover and 1.5 or 2 mm on the slide. To aid you, use a straight bar as a guide to support your hand, if you need this. Leave the sealer to dry thoroughly and paint another layer over it. Practice it. With practice it is not so difficult.

Don't forget to label your preparations and to make a cross reference to a separate catalogue of your filed slides. Commentaries: The objects mounted in Karo without additional operations have the appearance of those mounted in glycerin despite the low difference between their RI's. But it is much easier to obtain a successful and long lasting preparation with Karo.

Made this way Karo mounts last a long time. Certainly much more than 10 years judging by the scarce published testimonies. Anyway, file them in horizontal trays, and give your collection a review every six month or at least every year.

If you don't have the special flat trays sold to file microscopical slides use your classical slide box.... but file them vertically. If you want to use pure fructose, search the supermarket in the sweeteners section. It should be there, but it is somewhat dependent on the dietary trends of the moment. If it has glucose as an added component, don't worry, it can be used faithfully.

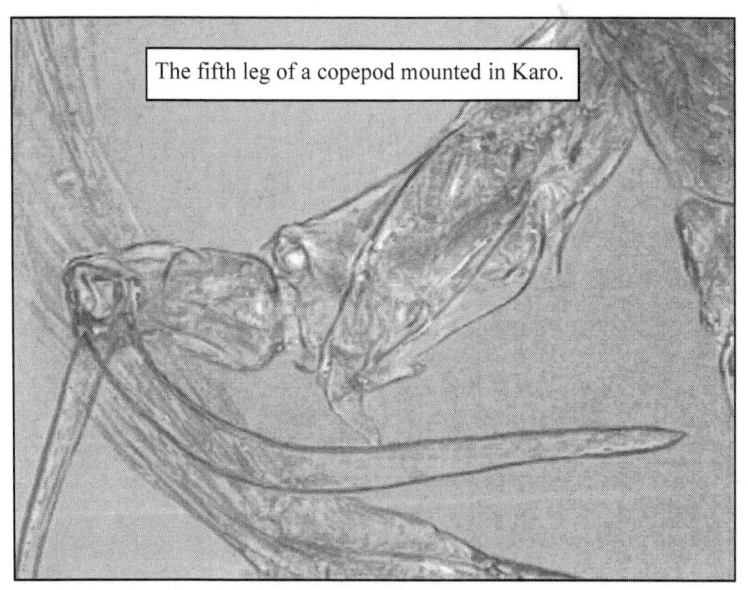

The fifth leg of a copepod mounted in Karo.

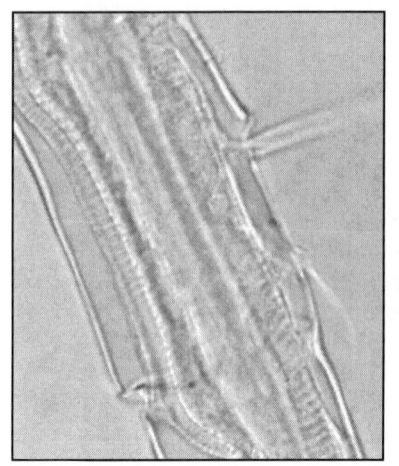

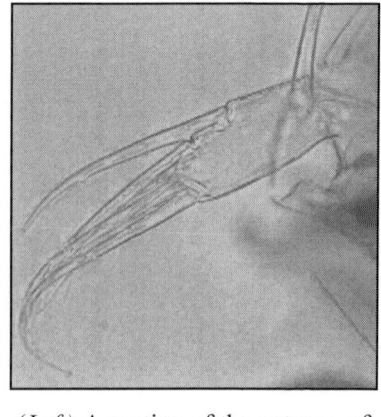

(*Left*) A portion of the antenna of a female in another specimen. You can see the striated muscles. One cell shows its nucleus. (Karo mount). (*Right*) A portion of another leg of the same copepod. (Karo mount).

Mounting other objects in Karo Mounting Media (KMM)

Objects that need clearing
If you try to mount somewhat dark chitinized subjects, you will probably like to make them more "transparent". Give them a preliminary treatment with lacto-glycerol (with previous steps through glycerol dilutions, of course, or using the Pantin method), and immerse in Karo after draining well, without another treatment.

Thick objects
Sometimes you will need to mount a thick object. Just start by including in the drop over the slide three or four supports of a height similar to the object height (hairs, paper, small pieces of broken coverslip and so on). Use Karo full strength or even a more concentrated solution obtained by evaporating the commercial one. Be generous with the quantity of mountant so there is a surplus surrounding the coverslip. It is best to leave the preparation to dry in the air. Be prepared for a very long process. As the solution dries, Karo retracts from the borders. If you need to, you must replenish the voids with concentrated solution. Use a very fine glass rod, or a needle, to supply the replenisher to the coverslip borders. When you see, after several weeks, only a little rate of retraction, and that the borders are only a little sticky, you must clean and seal the borders of the coverslip. Apply two layers of sealant. File horizontally and review every six months or more frequently until you are convinced that the preparation is set. Most of these preparations

must be studied only with the low power objective, or under a stereo microscope (e.g. mosquito larva).

Fragile or tender objects
Many objects suitable for this mountant could be more difficult to manipulate. Microalgae, filamentous ones, hand made sections (of marine algae, or of stems, or leaves of other plants), entomostraca and other small arthropoda, or some worms and so on. Methods to fix and stain some of these materials will be explained in future articles in this series. Assuming that you have your materials ready to mount gathered in a capsule of water. You must try one of these two techniques:

1.- Use with Karo the same sequence of steps we used with glycerin. Start with very diluted media between 2 and 5%, and increase in small steps of 5 or 10%. Be patient, the fructose solution is dense enough to cause plasmolysis in the tissues of the mounted critters.

2.- It is indeed possible that your material is not suitable to support this series of transfers. In which case you must select a very large capsule and pour in it a 2%, 5%, 10% or 20% solution of Karo depending on your subjects. Include your material and leave it in a dust protected environment. Allow the evaporation of water, spontaneous or helped by gentle heating. When you judge by the volume reduction that the subjects are in pure or at least very concentrated Karo you can proceed to the final mounting steps.

Don't think that these are undesirable characteristics of Karo. First of all remember that your goal is to work with a safe mountant. And you must know that with all mounting media you need to take similar precautions for these kinds of materials. And with some, as for the resinous mountants (including the famous Canada balsam) it is many times impossible even by using the most careful methodology to obtain the quality preparations that the aqueous media can provide.

Nail Polish Mountant
Some months ago I presented nail polish as a useful and (more or less) safe mountant. For further information, I only wish to say here that you must try the consistency of your product and to dilute it, if necessary, with a little acetone. The mountant must run easily, otherwise the materials you try to mount can be displaced by the movement of the too thick liquid as it reaches the borders of the coverslip.

You have two ways to incorporate your materials. You can mount "dry objects" in the NPM directly from alcohol 96%. If you start from water use 30%, 50% and 75% steps to reach the full concentration. If your subjects allow this, this is the preferred method. Or if your subjects are difficult to impregnate, you can make two additional intermediate steps, one in half alcohol/half acetone, and other in pure acetone. Alcohol and acetone are easily miscible with each other and with the NPM. Depending on the size and nature of the subjects 2 to 10 minutes is enough time to prepare the mount.

The slides dry in no more than an hour. It is the fastest of the mountants I discuss here. And gives, without pain, very clear images, most similar to the materials cleared in lactoglycerol at half strength. Try Nail Polish as a mountant. It is rewarding.

Commentaries: *Treat the materials from glycerin with care. Drops of glycerin can be included in the NPM and persist in the mountant as less refractive discs. If they are not over the object it can be only an aesthetic matter, but in my recent experience, not all "nail polishes" accept glycerol. Try your local brands.*

IMAGE GALLERY OF NPM MOUNTED MATERIALS

1 - The "TEST OBJECT" mounted in NPM four months ago. (Objective X40).

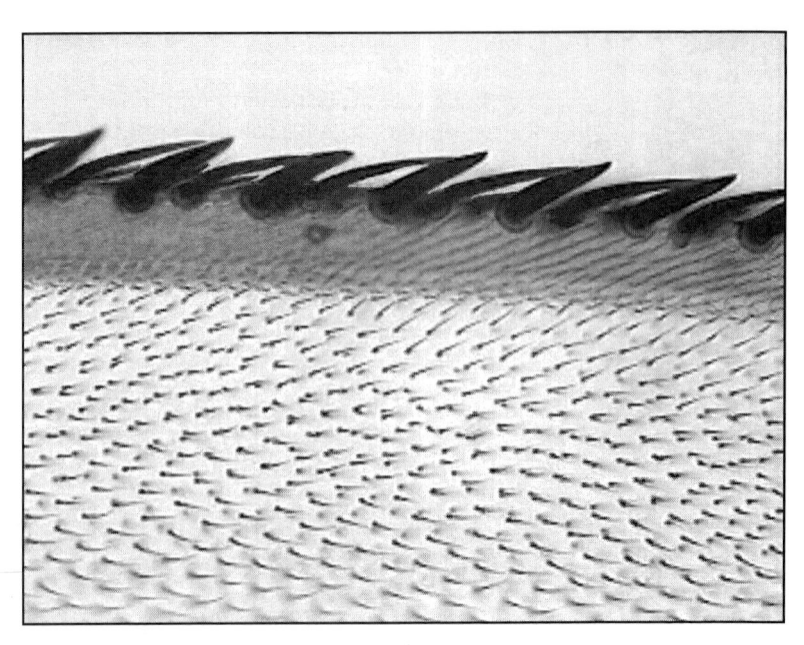

1

2 - THE ARTICULATION OF A MOSQUITO WING (From the same slide used for the chapter on NPM.)

3.- EPITHELIUM FROM THE SAME KIND OF LEAF USED WITH KARO. X10.

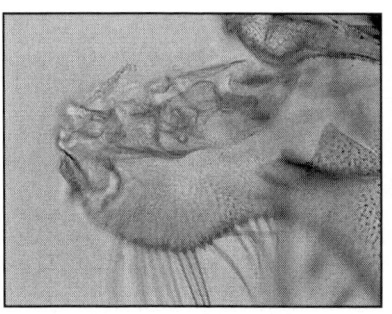

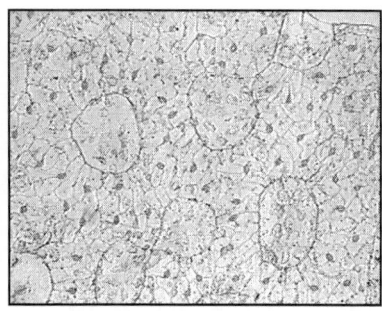

4 - Right. The same subject as in [3] at x100. Note in both pictures the differences with the Karo mount in cell definition, and the total decolouration of the chloroplasts, which is almost immediate.

5 - Below. The 5th leg of a diaptomid copepod x40.

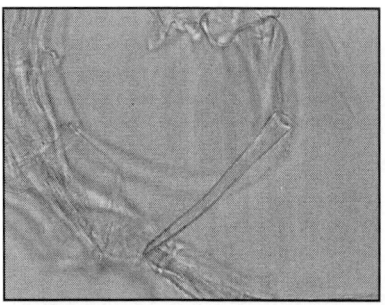

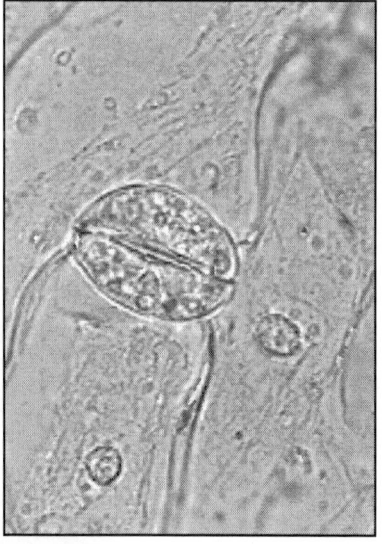

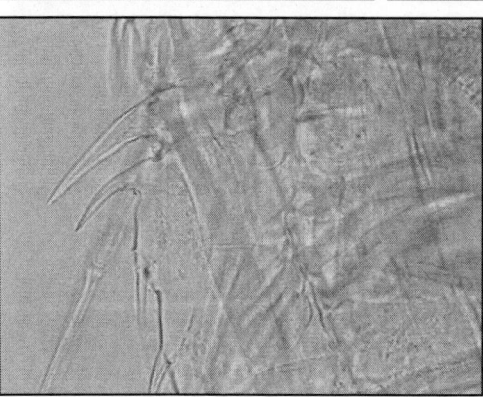

6 - Left. Border of a portion of the antenna of a male x40.

7 - Next page. The same subject at x100. Note the nucleus of the inside tissues.

8 - Next page. Through the exoskeleton can be seen some muscles of the buccal pieces and the cells of the cerebroid ganglion.

7　　　　　　　　　　　　8

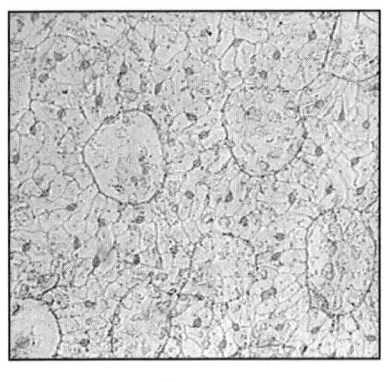

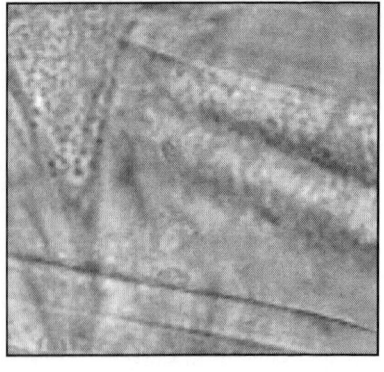

OTHER SYNTHETIC MOUNTANTS
Cyanoacrylate.- Loctite™, is one strong glass bond proposed as a mountant in a previous issue of Micscape. It seems to have some similar properties with NPM. I have tried a different trade mark of cyanoacrylate which I can obtain in Durango. I find it excessively expensive, and furthermore after one year it has crystallized totally on the slides and in its tube also. So it is possible that different products behave in a very different way. Try the local products.

Recently Gordon Couger kindly sent me a note and links about a safe synthetic product which due to its high RI = 1.58 is recommended as a mountant for diatoms, coccolithophorids, sponge spicules, and similar subjects. The product is a plastic resin, that you can get from Edmund Scientific Co., that must be cured with UV radiation (i.e. sunlight) named NORLAND 61. This is the relevant link http://*www.couger.com/microscope/norland.txt*

USING A COMMERCIAL PVA GLUE AS A MOUNTANT
CPG- Polyvinyl Alcohol.- All the proposed PVA formulae for microscopic techniques are prepared from a high molecular weight product and are always mixed with lactic acid. PVA mountants have been proposed to mount fungi, and acarii. They are certainly useful for all microarthropoda and many other plant materials. To get small quantities of the adequate PVA is the problem. I have found a commercial clear syrupy solution of PVA sold as a paper glue (Itoya, O'GLUE JR.) in 1 oz. containers. So I tried it direct from the bottle.

I mounted my test wing with exactly the same technique I used with Karo. Using water, 20% and 50% PVA/water solutions as intermediaries. The medium dries hard in not more than a couple of hours, and is very easy to manipulate, except for the easy inclusion of big air bubbles. I get rid of them by pushing them to one margin of the drop on the slide, cutting off the area with bubbles with a needle set straight over the glass, and absorbing the liquid outside the needle with a piece of paper towel.

Commentaries: *Its refractive index is inferior to all the previously reviewed mountants, so be wise when selecting subjects to be mounted in O'GLUE, if you find it, or a similar product you may like to share your experiences in Micscape. When the medium has set, cut off the excess along the border of the coverslip, it is very easy to peel the dry media from the slide. I seal the coverslip with nail polish. In my experience, if you don't, air bubbles can develop in the interior of the preparation, probably by evaporation of solvent.*

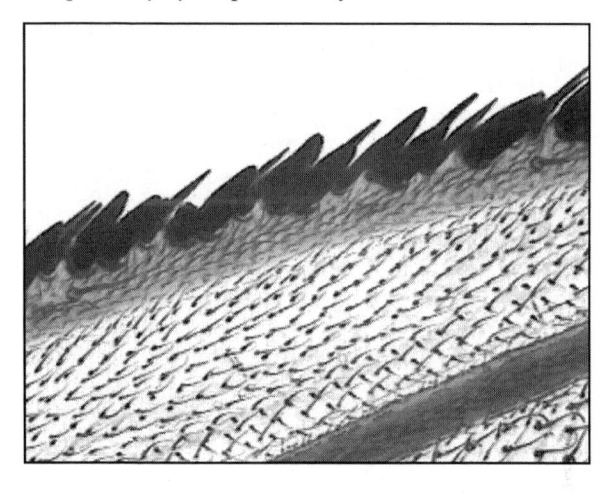

The "test object" mounted in CPG (polyvinylalcohol).
Above - wing border. Below - detail.

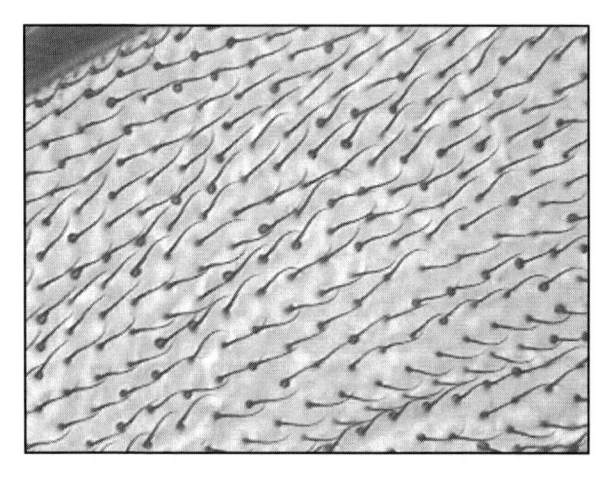

Chapter APPENDIX

Damar resin is an exudate of certain trees and has a long history in microscopy. It is used generally as a solution in xylene, and some authors think that in this form it has similar properties to those of Canada balsam. A group studying gregarines recommend it enthusiastically, as a long lasting media (the same as balsam at least) that preserves well the carmine staining (and possibly the other standard histological stains). Robert Constantine, an Australian e-correspondent, tells me that he currently uses Damar to mount the Australian hydracarina he is studying, but, as the xylene solution of Damar has a higher RI than he likes, he adds eucalyptus oil to the xylene Damar solution to successfully correct this problem.

Langeron in his 1954 "Precis de microscopie" says it is soluble in xylene, toluene and benzene. They are three dangerous products. Benzene is reputedly carcinogenic, and like toluene and xylene (although not carcinogenic) causes serious damage to the nasal epithelium, eyes, and brain cells, if they are inhaled frequently. Xylene is the less dangerous. Beside this, as a xylene solution, mounting in Damar requires the same careful steps that balsam does, with changes through increasing levels of alcohol and other solvents to attain a good dehydration. But it also happens that Damar is soluble in the so-called "spirit of turpentine" or "oil of turpentine" or "essence of turpentine". Turpentine is itself a resin exuded from coniferous trees, its distillation gives the liquid solvent.

Damar and essence of turpentine are common products sold in artists' supply stores; both are cheap and sold over the counter, in small quantities. Probably, (although I have no experience), the essence of turpentine doesn't preserve the histological stains.

But, allowing for this possible difference, we have a cheap, easy to find, useful balsam substitute....that receives objects directly from 96% alcohol, dispensing with the use of all the dangerous solvents which balsam needs. I recommend it only for grown-ups. Because the essence of turpentine, although safer than all the other solvents, is not totally safe. To a much less degree it produces the same problems as xylene. But you don't need to deal with the vapours for more than a few minutes when you prepare the solution. And you can do this with all the recommended safety precautions such as working in a well ventilated area, not to inhale directly the essence fumes, and be careful to ensure the flasks are always well stoppered. The other reason to ask for a grown-up to manipulate Damar is that it needs to be melted over heat, and to be mixed hot with the flammable essence.

This is the technique: Select clear and clean "drops" (also called "stones") of the resin. Put one part of Damar in a stainless steel vessel of sufficient capacity and heat gently (e.g. on an electric heater, not a naked flame) stirring with a glass bar until the resin is completely melted. Remove, and at some distance from the heat source, add two or three parts of essence, continuously stirring until well dissolved. The cold liquid must be of the consistency of a common nail polish. If in spite of all your precautions the solution has many suspended solids, the only solution is to add more essence, and to filter the solution, losing some medium of course. As the filtered liquid would be thinner than you need, you must evaporate the excess of solvent, leaving the flask open for some days, protected from dirt, where the solvent vapour does not harm anybody. (Robert Constantine tells me that he simply leaves his xylene solution to stand for some days then decants from the dirt and transfers the supernatant clear liquid to a new flask.)

To mount your subjects have them fixed, and dehydrate them through 30, 50, 75 and 96% alcohol. Put the object(s) on a slide, and absorb with a piece of paper towel most of the alcohol you can. With a glass rod put a drop of the resin over the object(s), arrange them as you wish, and cover with the coverslip. Within hours the resin replaces the alcohol and clears the mounted objects. Leave aside for some days to set.

As you see the only dangerous steps are in the preparation of the Damar solution. The post preparation use is as safe as all the other mountants discussed in my articles. So there is hope for the younger microscopists.

They only need to have the assistance of an adult to prepare the mix.

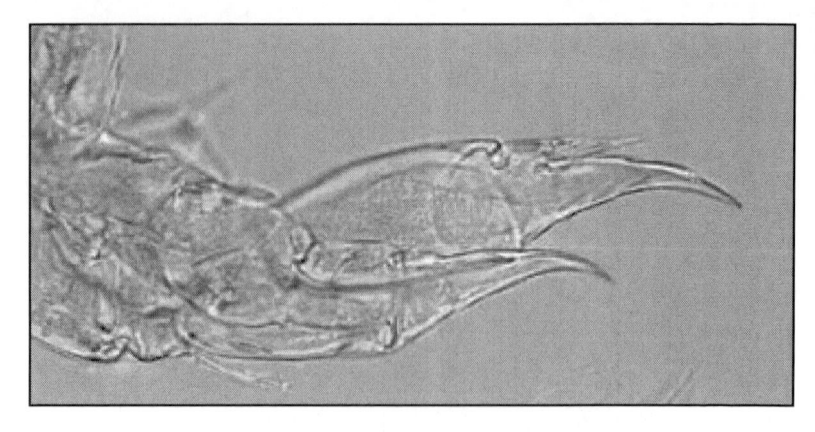

MOUNTING MICROSCOPIC SUBJECTS.
Chapter 3 - The Mixed Formulae

All pictures originally captured at 640 x 480 pixels. Many of them are amalgamated with CombineZ (a focus stacking programme). The title image is a picture of the fifth leg of a female Diaptomus mounted in fructo-glycerol medium.

PRELUDE
This chapter is in four sections which are interlinked.
Part 3a below: Introduction fructoglycerol and modified Brun's medium as mountants.
Part 3b: PVA-lactic acid and PVA-glycerol mountants.
Part 3c Gum arabic fructose, glycerinated gum and glycerinated lactic acid mountants.

I have shown previously what you can do with the principle components of the safer mountants. (Glycerol, Karo or Fructose, PVA, NPM and even Damar). But searching for more alternatives and to accomplish some special tasks, microscopists have designed mountants made of multiple ingredients, which complement each other. I want to review two PVA derivative mountants, fructose syrup derivatives, with added gum arabic and glycerol, and also the useful glycerol derivatives with added gelatin or PVA.

I've tried to always present at least one original formula, for those that have the opportunity to obtain the right reagents, and have also the equipment to prepare the "professional" mountant. All are safe formulations.

For the youngest of you, and those that need to work in less favourable environments, I present functionally similar formulations with the products you can gather from the supermarket, the normal drugstores, the old style pharmacies, the suppliers to college laboratories, the hardware stores, or the artist's suppliers.

I think that my own experiments were successful and encourage you to be creative and to look around you for the required reagents.

NOTE

When I say water you must understand that ideally it must be distilled water so as not to introduce undesirable chemicals into the formulae. You probably don't have a distillation unit in your house, and to install one is neither easy nor totally safe. Most of the time you can buy "distilled water" in the supermarket, sold as water for the steam irons, or for car batteries. Possibly this is not really distilled, but demineralised. The difference doesn't matter for amateur uses. If, in spite of all your efforts you can't obtain these qualities there are in many countries bottled potable water that is acceptable. Please! Never use the water from the municipal supply.

ANOTHER NOTE

Even when using safe chemicals you must be careful. Protect your clothes with laboratory aprons; and tables and floor, if they aren't exclusively for laboratory use, with thick plastic sheets. Never use the kitchen ware and measuring tools. Have a dedicated set of equipment for your own laboratory work. You really only need a few essentials. Take special care when you need to use heat to make diluted solutions. Use a low flame, or preferably an electric mantle, use safety glassware, a long glass stirring rod, some insulated matting, or a piece of wood to prevent the contact of very hot glass with cold surfaces, and never put your face and eyes over an open heated vessel. There can be unexpected bubbles and splashes and even the hot vapours can be dangerous. Have your chemicals well stoppered and labelled, and read the label twice before opening the bottle to use the content. Return the stopper to the bottle immediately after use.

ABOUT THE IMAGES

I try to give you the most faithful images I can, showing the aspect with which the mounted objects appear under my microscope. It is really difficult because my camera has a resolution of only 640 x 480 and each medium has its own RI, which requires some adjustment to the illumination. All this is complicated by the use of CombineZ which is needed to render the three-dimensionality of many subjects. I diversified the test subjects to better assess the applicability of each mounting media, but this has increased the number of images, without increasing the variety. You may finish being bored of wings, and epithelia, but I think it is worse not to give you an idea of what any media can do.

CHAPTER 3A: Formulae Derived From Fructose FG—Fructose-Glycerol Medium

Brun's medium was formulated to use glucose as the "sugar" ingredient. I have modified the formula to replace the glucose with fructose (in my case, Karo syrup).

Fructose at a 76% concentration	24 ml
Glycerol	3 ml
Antiseptic (I use Listerine)	3 ml

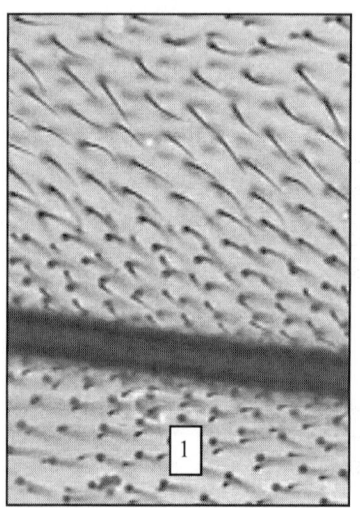

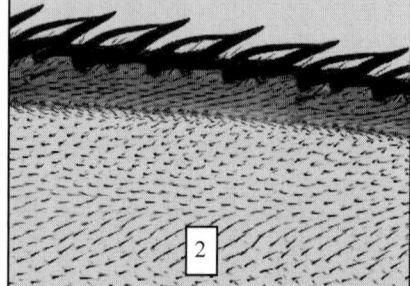

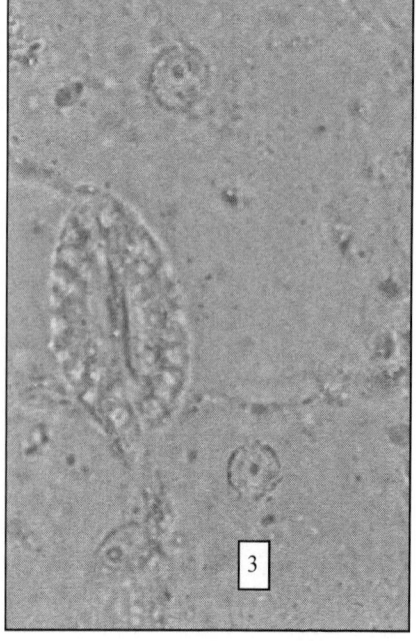

The first picture [1] shows the surface of a fly wing, and the second [2] the border of the same. The picture [3] is from the epithelium of the underside of a leaf, showing one stomata and the nuclei of two marginal cells. The fourth (opposite page) is the compound image of the 5th leg from a female of a species of *Diaptomus*, a genus of copepod (crustaceans). The sixth is the ovisac (a sac in which the female carries the eggs). All were taken of slides mounted in FG. The wings were stored in alcohol 70%. The epithelia was fixed for

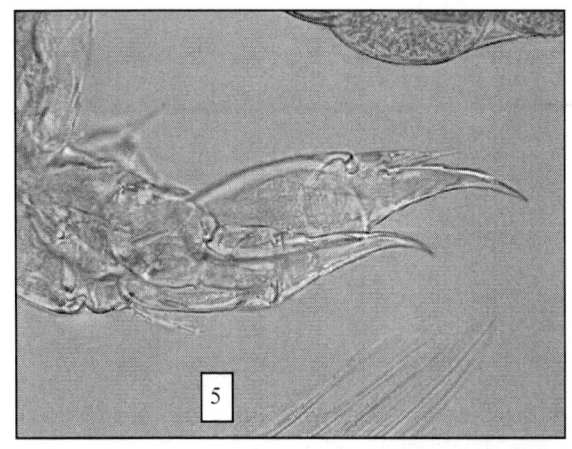

24 hrs in AFA (a fixative composed of Alcohol, Formalin, and Acetic acid) and the picture taken with the 100x objective (1000x), The copepods were fixed more than a month ago in lactocupric, a fixative that I will discuss in a future article, and this specimen was washed in water for ten minutes and directly transferred to the mounting media. As we will see in other images, this is responsible for the collapse of the eggs. Some intermediate steps through glycerinated water can prevent this.

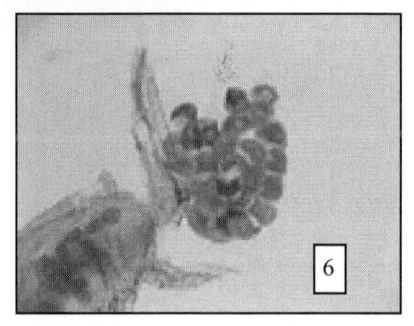

A modification that is also working for me is the addition of lactic acid as a clearing agent

FGL - Modified Brun's medium

Fructose (76% concentration)	21 ml
Glycerol	3 ml
Lactic acid	3 ml
Antiseptic	3 ml

These two media, give good results with the test objects but take longer to dry. I estimate the drops so they don't exceed the coverslip, and twenty four hours later I seal, with the same care as with a glycerin mountant. If not well sealed the coverslip continues to be "slippery" for more than a week.

The following images show examples. The first two *[wing]* & *[epit]* are similar to the FG pictures. The Diaptomus female in the third picture *[ovisac]* has adhered the ovisac, and also two spermatophores (sperm packets). Comments as in the FG case. The last three pictures are from a mosquito larva, fixed in alcohol 70% and first cleared in lactoglycerol, before being mounted in FGL. The head shows clearly the interior structure with the brain lobules, the nerves connecting to the eyes, several muscles and other sensitive organs. The ventral side of the head shows part of the buccal armature. The air tube is part of the caudal complex I depicted and labelled in the first article; you may wish to reread as a reminder of the anatomical features.

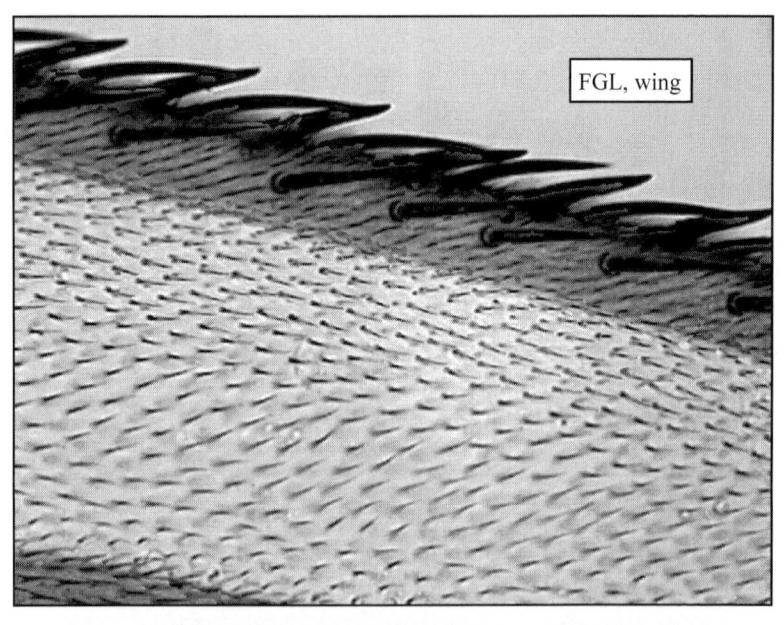

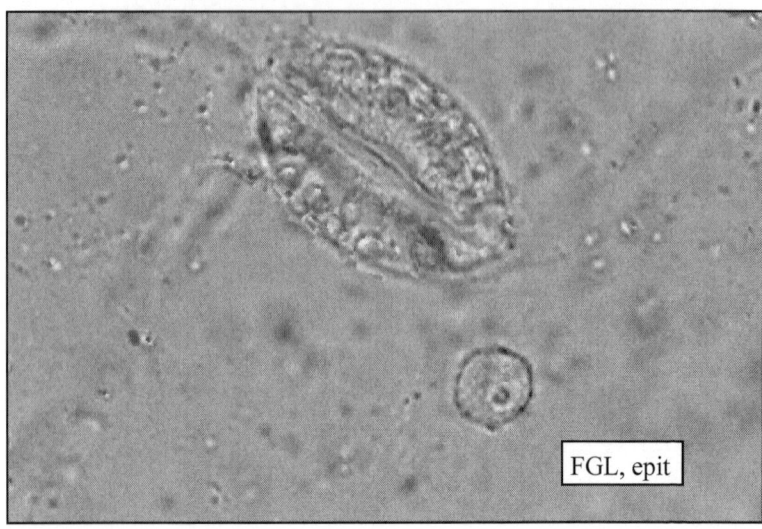

FGL, wing

FGL, epit

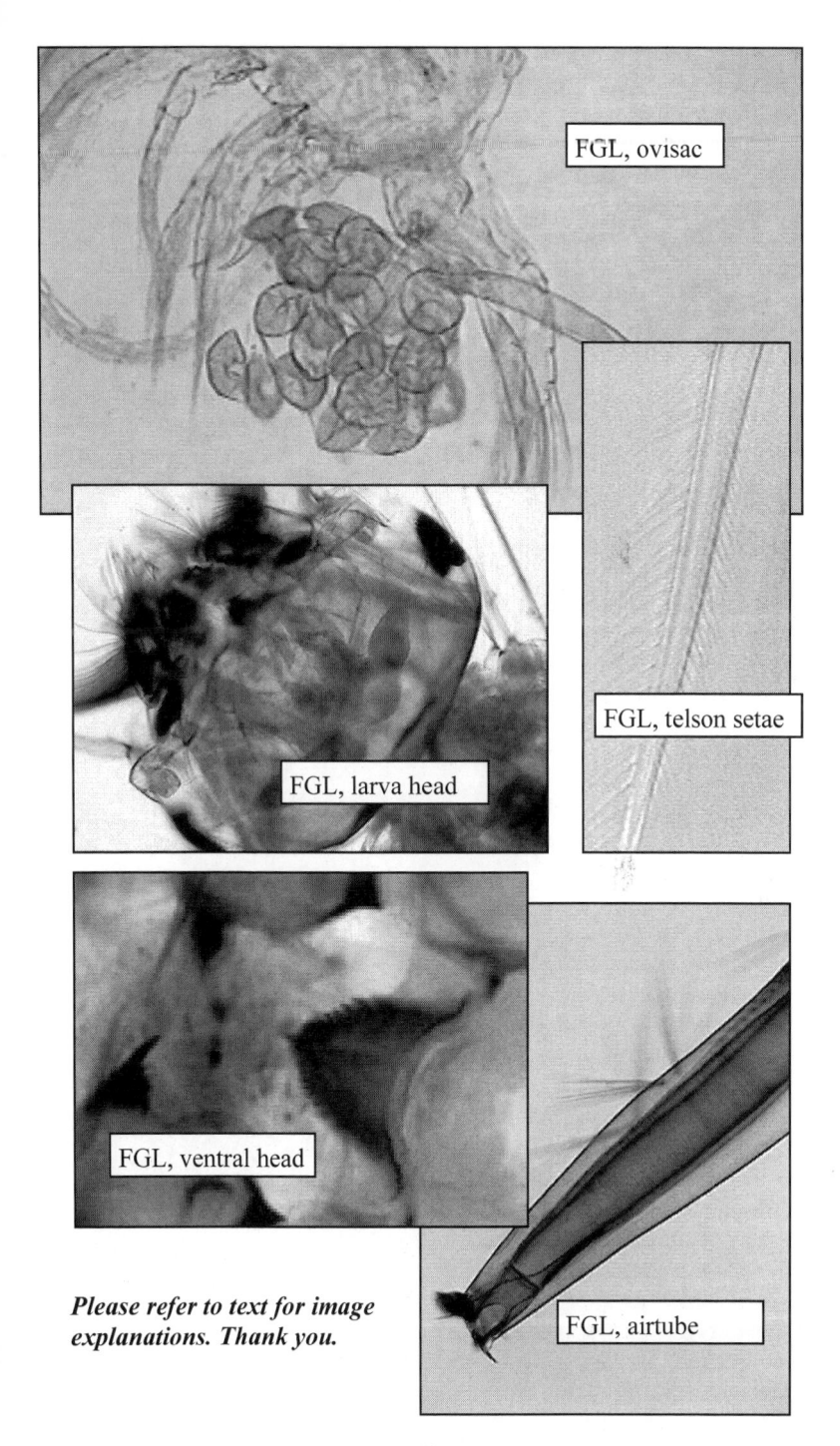

Please refer to text for image explanations. Thank you.

CHAPTER 3B: PVA-lactic Acid And PVA-glycerol Mountants

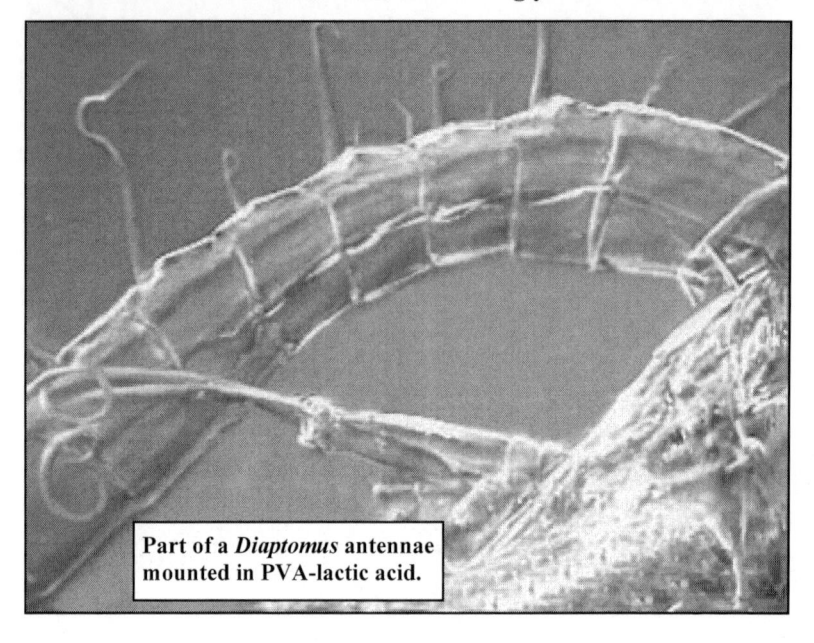

Part of a *Diaptomus* antennae mounted in PVA-lactic acid.

PVA-L.- Polyvinyl alcohol-Lactic acid medium

Lactic acid is not a forbidden substance. It is a clear liquid which has a mild odour and is pleasing to work with. As we saw when I presented lactoglycerol it is considered a powerful clearing agent coming only third to chloral hydrate and phenol. It is also an expensive product of very extensive industrial use, and more or less difficult to get in small quantities. But try old drugstores and suppliers to school laboratories, surely you can get the quantities you need at a reasonable price, as I did.

Probably the first PVA-lactic acid mounting media was the one published by Omar et. al. in 1978. You can prepare your own PVA-lactic mountant, with an RI of approx. 1.39. One professional formula I obtained from several sites on the Web is this:

PVA................................	16.6 g
Water...............................	100 ml
Lactic acid........................	100 ml
Glycerin...........................	5 ml

Dissolve PVA in water, add the lactic acid while mixing vigorously. Add the glycerin and leave for 24 hours before use. My (amateur style) is of this formulation:

O'Glue (See Part 2).......... 30 ml
Lactic acid..................... 15 ml
Glycerin....................... 1.5 ml

It has a very good consistency. PVA-L is a universal mountant. Suitable subjects include small arthropods, parts of the same, micro fungi, some algae, some botanical preparations. You can transfer the subject directly from water, alcohol or glycerol to the PVA-lactic media, or if the objects are really dark, you can use a preliminary clearing bath (2 parts lactic acid:1 part glycerin), then transfer your subjects to glycerin. Dilute with water if it is convenient. Another clearing medium could be simply lactic acid at a 1% or 5% strength. Experiment to find the best medium and the time your material needs to be cleared. Some subjects clear in minutes, others require as long as several days. Mount as above.

My "lactic" formula dries very rapidly. In some hours the media will set. As you will see the margins harden enough to clean up as I described for the CPG (commercial PVA glue) medium and seal the preparation with a double layer of nail polish. Some references state that even so sealed the PVA formulas can evaporate solvent, dry out and after some months peel off the glass. I suspect that NPM (nail polish mountant) can have the same behaviour.

It is recommended to seal with a really hard sealant. In the USA the Red Glyptal, a varnish to protect electric machinery (also used by palaeontologists to protect fossil bones) is recommended. I have still not found a substitute in Durango, and must use some automotive paints, hoping this helps.

Examples

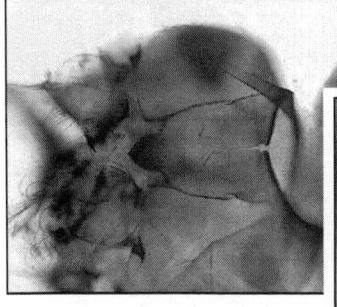

PVA-L, mosquito larva head ventral [Left].

PVAL, mosquito larva buccal armature [Right]

PVA-L, mosquito larva pecten airtube

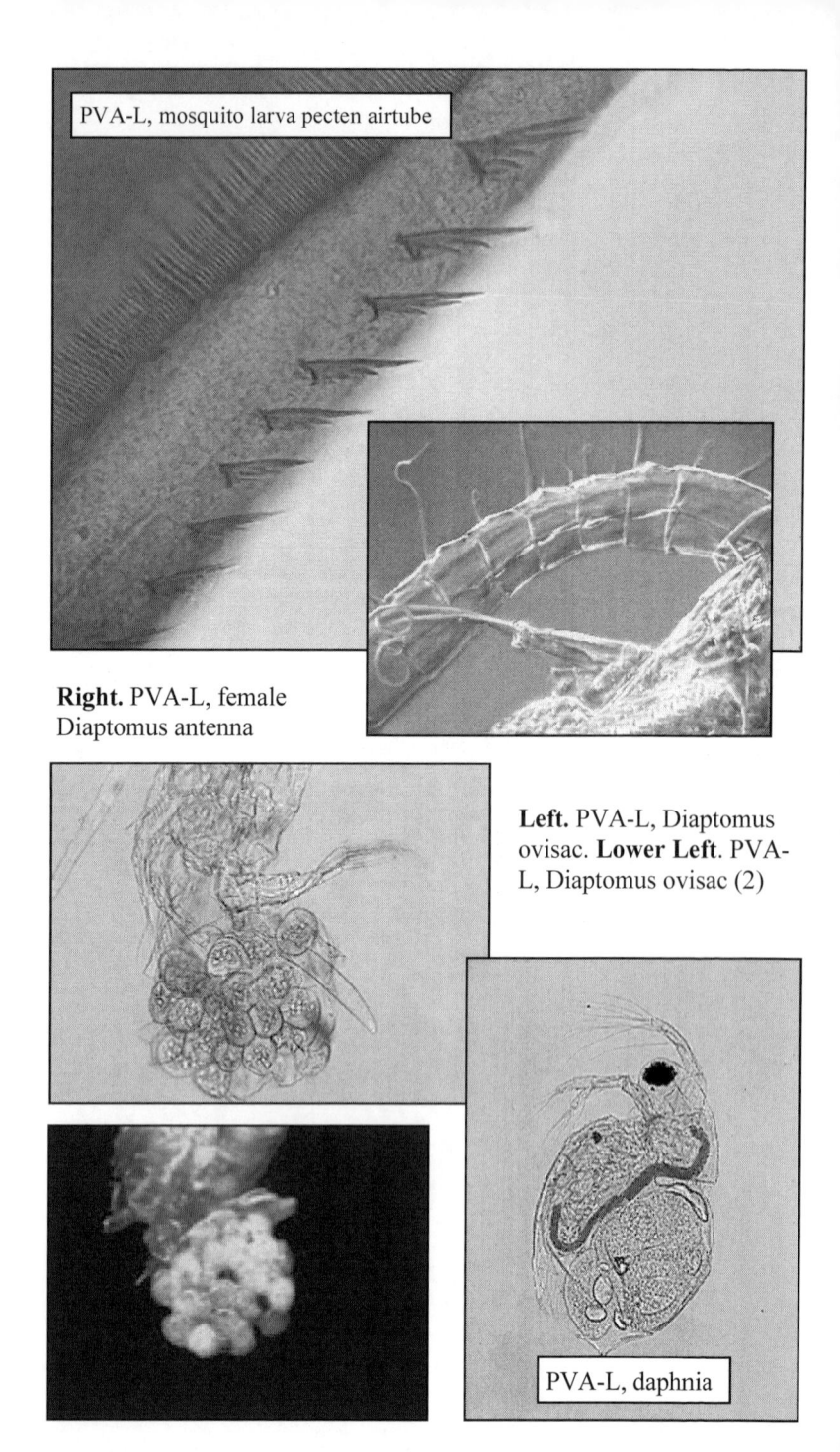

Right. PVA-L, female Diaptomus antenna

Left. PVA-L, Diaptomus ovisac. **Lower Left.** PVA-L, Diaptomus ovisac (2)

PVA-L, daphnia

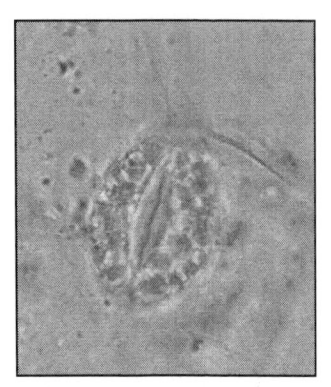

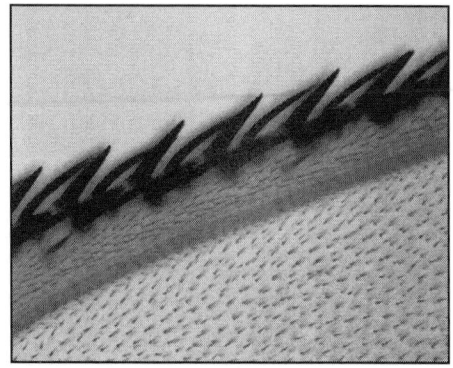

Left. PVA-L, epithelial cell of leaf underside
Right. PVA-L, fly wing-1
Below. PVA-L, fly wing-2

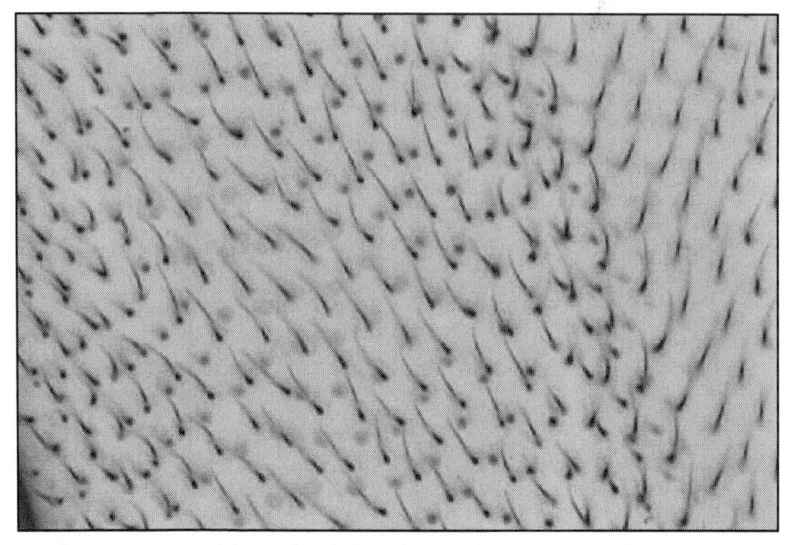

A few comments about 3 of the above pictures. The diaptomid first antenna is portrayed in my domestic version of the scanning microscope; (Rheinberg plus a vertically displaced central stop somewhat out of centre). The copepods were washed from the lactocupric fixative with water and mounted directly to the PVA-L. In spite of this, the eggs don't show the collapse that they suffer with FG and FGL. The ovisac in darkfield was photographed using the same method as for the antenna, but with the central dark stop well centered and adjusted, and I use a Rheinberg filter of another colour. (Well... as the black field doesn't register so black as I'd like it in the original picture, I made a substitution with the aid of Photo Paint.)

Contrary to the CPG, of a lower RI, this lactic formula I prepared has a similar behaviour to NPM (nail polish mountant) and gives similar results in the short time. With more time, (one or two weeks) it almost clears the internal tissues and gives a neat view of the chitinized structures. In several weeks the extent of clearing makes it more difficult to discern the thinner structures like spines and setae. You can try to stain the chitinous skeletons by treating them, before mounting, with lactic acid coloured with a few drops of methylene blue. (I picked up this trick from the web, but I have not tried it until now.) Or as a last resort use good oblique illumination or phase contrast.

So you must select very well the materials you mount in this kind of mountant. Or try the solution used by Robert Constantine for his Damar mountant. See also my commentaries to the Glycerinated Lactic Gum (to be published in Part 3d).

The literature on PVA mountants has many enthusiastic appraisals and also some totally dismissing opinions. The late G. Ramazzotti, in his monograph about the Tardigrada says that many times (as I have experienced with the O'Glue) the polyvinyl-lactophenol formula he used developed voids under the cover slide, without any known reason. He also states that on the contrary, he had preparations that lasted many years without faults.

I think that if one can obtain a PVA of the recommended density (24 – 32 centipoises), the professional formula is a safe mountant, easy to mix, that merits more additional experimentation. But if you are unable to get it, browse through the art or the office supply houses, as I did. Try the PVA paper glues, they deserve a try.

Some time ago I started to question why the PVA was restricted by professional microscopists to the lactophenol formulas (now additionally restricted to lactoglycerol formulas). So I tried with relative success the CPG adventure (see part 2). Now I propose you indulge in the heresy and design a PVA based media of mild clearing action, a lot less acidic, that could be of a more general application (including those little tardigrada, with his delicate calcified pieces). My own experimental version is below which I've put on the trial to monitor its behaviour for the next few months.

PVA-G.- Polyvinyl Alcohol-Glycerol medium

    O'Glue...................... 10 ml
    Borax Water*...............4 ml
    Glycerol..................... 6 ml

*Saturate water with granulated borax (>6 g of borax /100 ml water). Use the supernatant liquor.

Editor's note added Nov 2004: The preparation method of PVA-G is quite critical as apparently the above ingredients can also make synthetic 'slime'. ***See Howard Webb's commentary after Walter's image examples below.***

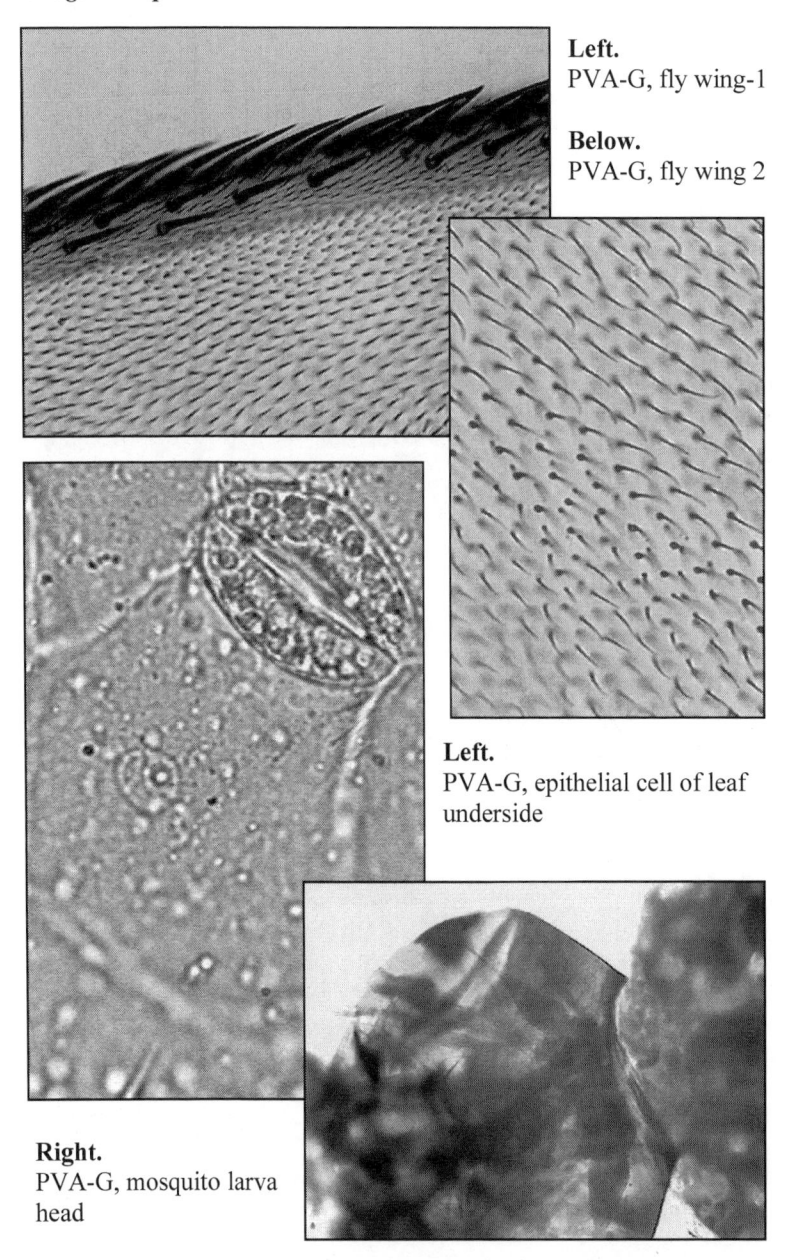

Left.
PVA-G, fly wing-1

Below.
PVA-G, fly wing 2

Left.
PVA-G, epithelial cell of leaf underside

Right.
PVA-G, mosquito larva head

PVA-G, mosquito larva pecten (1000x)

Postscript. **Polyvinyl Alcohol with glycerol. by Howard Webb (St. Louis, MO, USA)**

The following are some personal comments on PVA-G from an April 2003 Micscape article by Howard Webb.
. *Walter Dioni has provided a good formula, but there were a few surprises when I attempted to follow his instructions. A bit of a "warning" may help others. I like using this mountant, and would encourage others to try it.*

Additional Information
My standard mountant has been jellied glycerine, but after the nice articles by Walter Dioni, I decided to try his polyvinyl alcohol and glycerol.

The supplies were all locally available for only a few dollars. O'Glue was readily available at my favourite art supply store. A small tube will make about four batches. Borax was a bit harder to find, not being at my usual grocery or hardware stores. It did however turn up as 20 Mule Team Boraxo at Wal-Mart, though a box is definitely over-kill for the amount needed. It took asking several pharmacies to find a small bottle of glycerine.

The quantity of ingredients specified in the formula all fit nicely into an empty plastic film canister, so I measured out some water and marked "fill lines" for the various ingredients.

What Walter failed to mention is that polyvinyl alcohol and borax are standard ingredients with another purpose - "slime" (try a web search for "polyvinyl alcohol borax"). And the amount of borax will change the consistency from runny to almost solid rubber (before adding glycerine).

When I first tried mixing the mountant, and added a saturated borax solution to the polyvinyl alcohol, the contents turned into a solid block of rubber. I thought I must have made a mistake (and started another batch). For the second batch I mixed the polyvinyl alcohol and glycerol. It lumped, then frothed when I shook it. I thought this must also be a mistake. I then poured glycerol over the first mixture (the rubber block), and tried chopping it in. In near despair, I set both aside in a warm place (on top of a fluorescent light ballast), and left them for the night. The next day, when I checked them, they had both turned into the desired, clear thick syrup.

When I go to make another batch (years from now with my current supply), I would add the PVA, glycerine and borax (in that order) to a container, then shake them all together, then gently warm it to remove any bubbles.

CHAPTER 3C: The Mixed Formulae - Gum Arabic Media

GUM ARABIC MEDIA
APT- von Apáthy original formula

Two references on the web claims an RI = 1.52. I think one is an error, the other a copy of the first. All my four old references state a credible 1.42 RI even with the fructose modification discussed later.

gum arabic............................ 50 g
refined sugar........................... 50 g
water...................................... 40 ml
antiseptic............................... 10 ml**

    The recipe states that you must use clean drops (stones) of gum arabic. Mix sugar and water and add the gum last. Put in a sealed jar. Completely dissolve it on a stove at 40°C (from a few hours to days) stirring only occasionally and slowly. Or use a water bath at 60°C for a more rapid dissolution. Take care not to include air bubbles. Finish adding the antiseptic. The antiseptic is needed because without it the medium develops fungal infections in less than 2 weeks. The amount of water can be raised to 90 ml.
    This is the general method to prepare the gum solutions recorded in the literature. If it is thicker than you want, or if you want to filter the liquid, dilute with enough water. After filtering you must concentrate the medium by evaporating the water on a stove.
    It is a safe and cheap formula. The problems are to get the drops of arabic gum, the difficulty and the time needed to dissolve it…and the frequently reported problem of sugar crystallization. When I was a biology student all the "lipid tissue" slides in our histology lab were stained with Sudan III, and mounted in Apáthy's medium. Many of the slides were more than five years old, yet I've never seen a crystallized one. But…

Examples.

Right.
APT, fly wing-1

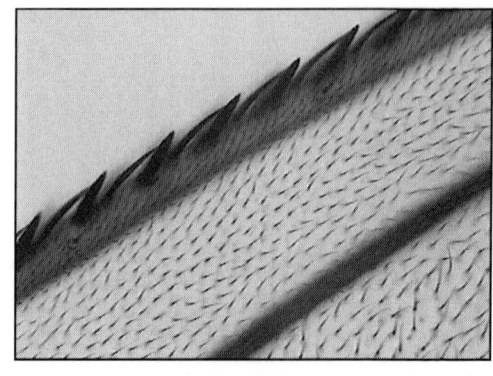

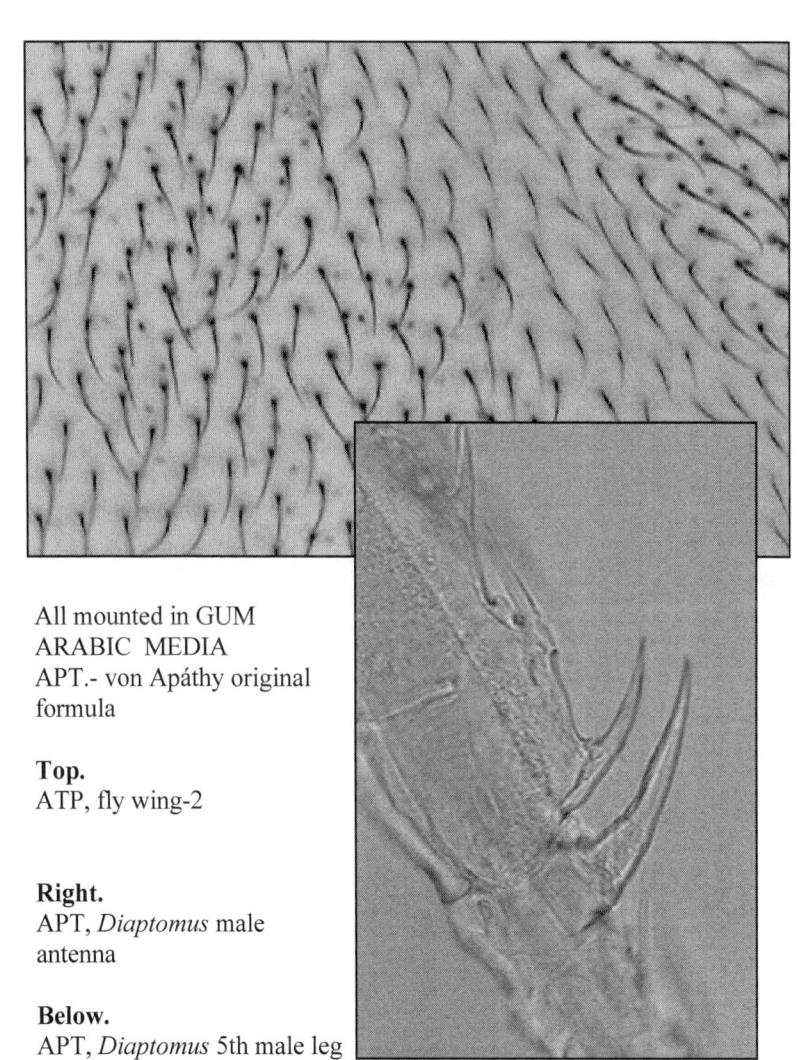

All mounted in GUM
ARABIC MEDIA
APT.- von Apáthy original
formula

Top.
ATP, fly wing-2

Right.
APT, *Diaptomus* male antenna

Below.
APT, *Diaptomus* 5th male leg

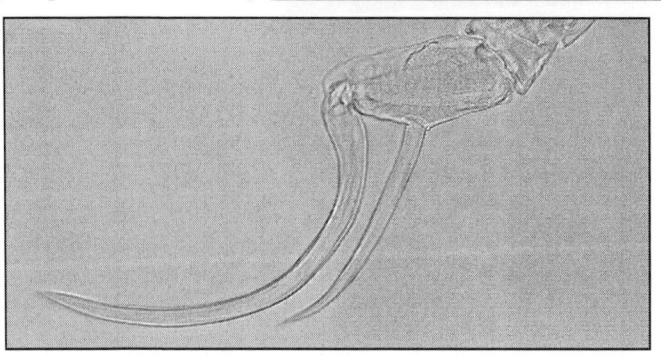

Left. APT, daphnia eggs

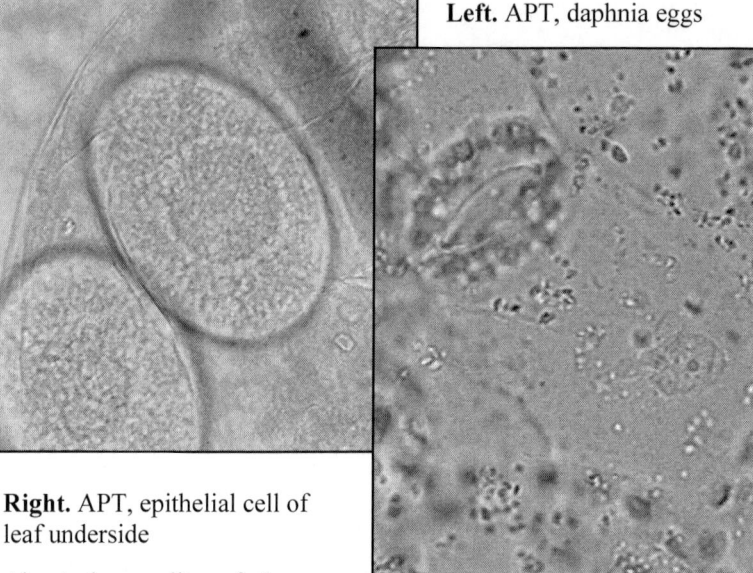

Right. APT, epithelial cell of leaf underside

About the quality of the gum based on my trials within the last hour! I insisted with my apothecary on the quality I required of the Arabic gum. He consulted with his suppliers and assured me that the grade he was offering was really Arabic gum, powdered, but pure. I remembered my old days as a young scholar when I bought the powdered Arabic gum and put it in water, only to finish with a sphere of moistened compacted powder enveloped by a shiny bubble of air, and all that work to break it and dissolve (over many hours) my paper adhesive.

I was near to refusing the offer when I remembered my "cocoa trick". My powdered cocoa, like the Arabic gum with the water, refuse to mix with my cold milk. But I have a trick, you know, I mix the cocoa with the sugar, and later I add the milk. And this works. So I bought some gum. Try this technique:

1) In a 30 or 50 ml flask, put 1 part powdered gum, and 1 part sucrose. Seal and agitate thoroughly to obtain a dry homogeneous mixture.

2) Slowly add portions of the water, mixed with the antiseptic. Any sugar crystals act as micro tunnels that carry the water down the mix. Carefully move the powder with a needle if it stops running.

3) When all is wetted (it took one minute) leave alone for a time (4-6 hours) the glutinous mass, or put the flask in a bath of water at

60°C (or a little higher).

4) Presto! You have your gum arabic solution made in five minutes of not so hard work, and all you need now is to allow some time for all the dirt to precipitate to the bottom (perhaps overnight).

5) Decant the supernatant liquid and adjust the density to suit your preferences. von Apáthy's medium isn't compatible with normal histological stains. The pH of the medium is near 4.0 (highly acidic) so stains fade or bleed into the medium. If you only mount unstained materials, or you intend to preserve your slide for only a few weeks, this is not a concern for you. But some modifications claim a better behaviour for the formula. Three alternative added chemicals were proposed: potassium acetate (30 to 50 g added to the formula), calcium chloride (10 g) or even sodium chloride, the common table salt, (10 g) It seems that these additions raise the pH to near 7.0

Mount using the technique proposed for Karo. It is not necessary to seal the cover slips. But I do.

** The original formulas for gum arabic or gelatin media always include an antiseptic: thymol crystals (one crystal), phenol (1 g), Mertiolate (10 ml), etc. All are banned as they are either difficult to obtain or are very toxic. Remember that this additive functions just as an antiseptic so you can even eliminate it without affecting the mountant's performance, and take the risk (most improbable) of a bacterial or fungal infection of your slide, as M. Cairn and J.M. Cavanihac have done. Modern alternatives could be benzalkonium chloride, Listerine, sorbitol, or potassium benzoate.

As Mertiolate contains mercury bichloride it was banned, and the suppliers changed the formula of the product. For a long time a solution of benzalkonium chloride has been sold under the trade mark name of Mertiolate. It was even coloured red as the original product was, which made it unusable as an antiseptic in this formula. Now one can find white reagents under the Mertiolate or benzalkonium chloride name. When you mix it in the formula you could be worried by the milky cloud it forms. Continue to stir carefully until the mixture becomes clear again.

Listerine is a mouth washing solution. Its formula is composed of several disinfectant ingredients that are also used as clearing agents in microscopy. The old formulation had the toxic phenol in its ingredients. It is eliminated in the modern formulation. The published composition is:

menthol..........................42.5 mg	ethanol........................22.7 ml
thymol........................…..63.9 mg	water..........................…77.3 ml
methyl salicylate...........66.0 mg	eucalyptol..................…...92.0 mg

The Listerine I get has a slight blue green colour (turquoise, says my wife), but in the used quantities this is not a problem at all.

LMM - Lillie medium.- RI = 1.43

I give you one of Lillie's modifications of von Apáthy's formula. If you use the powdered gum arabic, also use the fructose as crystals. Add fructose and potassium acetate to the powdered gum. Mix well; add the water and the antiseptic.

gum arabic (powder)……………..50 g
fructose (crystals)…………………50 g
potassium acetate…………………50 g
water…………………………… 90 ml
antiseptic…………………….. 10 ml

It has a good refractive index and a pH of 6.7. Most commentaries are as for upper formulas.

MY GUM ARABIC FORMULAE

As I can't find drops of gum arabic (and I had not have yet remembered my cocoa trick) I made another trip to the art supply store, and returned with a medium syrupy solution of it as sold for artists to mix their watercolours and for finishing their art works. These formulae are provided as examples for you to develop your own. Some of them do not contain sugar, so you can't do the old trick. You need the commercial solution.

GAF.- Gum Arabic-Fructose medium (RI probably around 1.42)

Gum arabic solution……………………………….30 ml*
(Commercial Karo "Clear" (for babies)) ……….30 ml **

*If the commercial solution contains excess liquid, concentrate it by evaporation.

** or use Larry Legg's fructose solution.

I omit the antiseptic because for obvious reasons the commercial solution must have one. Those concerned about bleeding of histological dyes can include 15 g of potassium acetate (or try the cheaper and ubiquitous sodium chloride…. 1 levelled teaspoonful) to raise the pH.

The fructose substitution for the sugar lowers the danger of crystallization. It is a worthwhile and safe precaution to seal the coverslip to stop evaporation.

Mount directly from water and as if it were pure Karo. File horizontally.

Examples

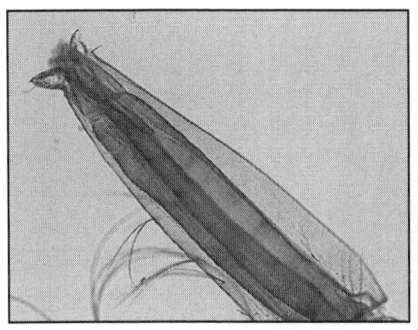

Above. GAF, airtube of mosquito larva

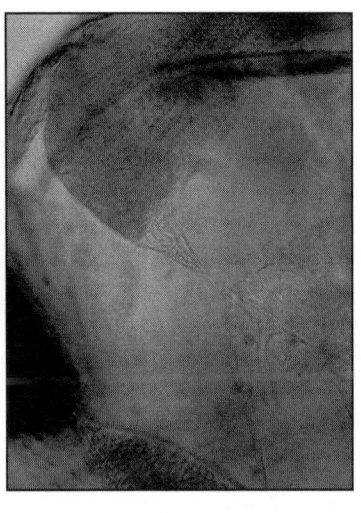

Right. GAF, mosquito larva head nerve ending

Left. GAF, gland in mosquito larva body

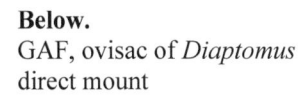

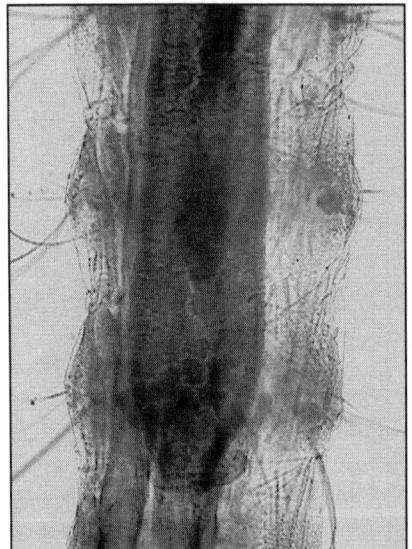

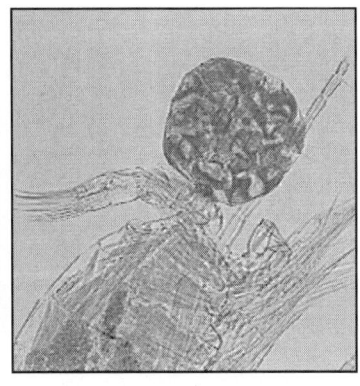

Below. GAF, ovisac of *Diaptomus* direct mount

Right.
GAF, ovisac of *Diaptomus* stepped mount

Below.
GAF, *Diaptomus* male antenna

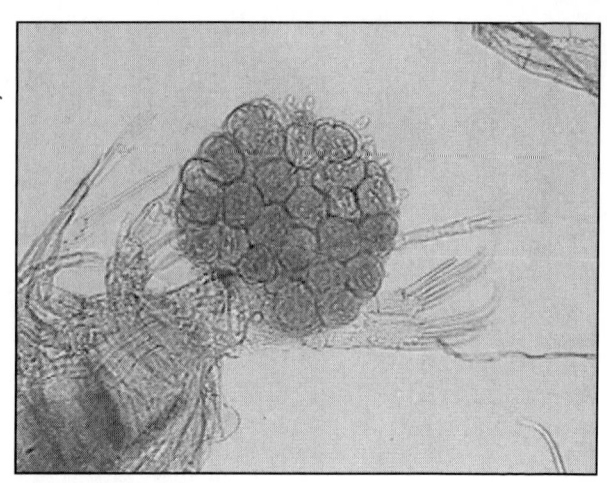

Below.
GAF, ovary of *Diaptomus*

Below.
GAF, epithelial cell of leaf underside

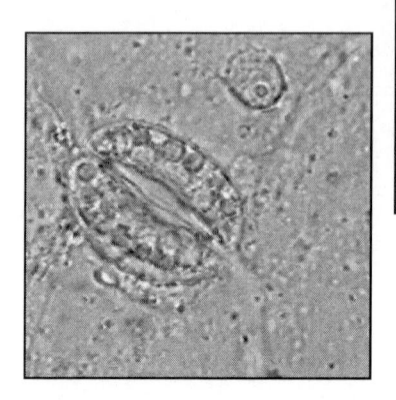

All mounted in GAF.- Gum Arabic -Fructose medium (RI probably around 1.42).

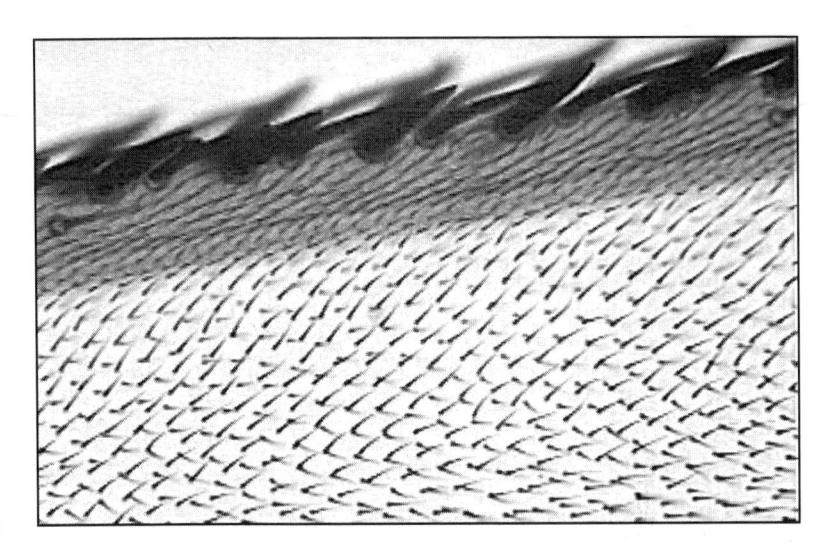

Above. GAF, fly wing

Through the skeleton of the head it is possible to discern what is probably the start of the "dendrite" of a nerve cell in contact with a sensitive cell. The other most interesting pictures are the *Diaptomus* ovisacs (one collapsed after being mounted directly from water, and the other normal, after being worked up through 3 dilutions of glycerin in water) and the cells of the ovary of one of the females, seen through the dorsal exoskeleton.

Commentaries: This is obviously my version of the Lillie's medium. The RI, and the appearance of the mounted subjects are good, and it dries fast (in my climate). As it is somewhat liquid (at least in my flask) carefully estimate the drop size to avoid excessive mountant overflow. I challenge you to adapt this formula using powdered arabic gum and fructose crystals.

GG.- Glycerinated Gum.- (RI about 1.44)
This is a formula in the style of those proposed by Farrant, Lillie, and Dahl.

gum arabic solution....................	30 ml
sugar..	30 g
glycerin....................................	30 ml
antiseptic.................................	10%

If you want, you can add sodium chloride (10 g) or 20 g of potassium acetate.

Examples

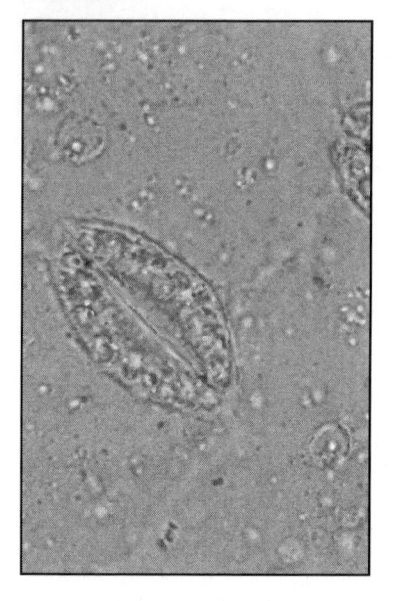

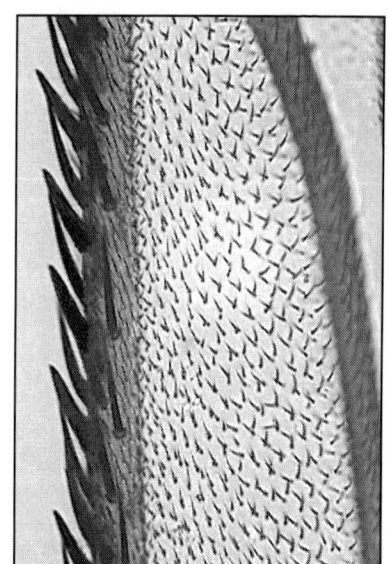

GG, epithelial cell of leaf underside

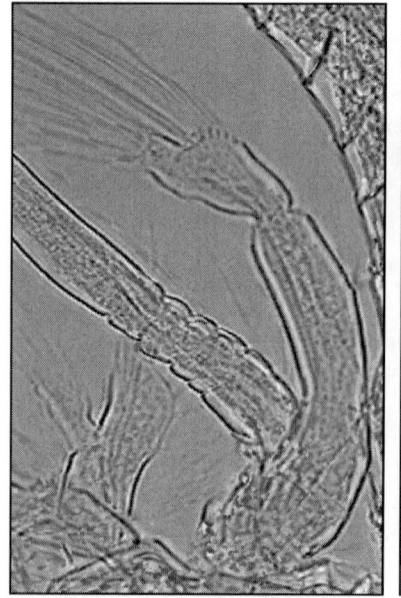

GG, fly wing-1

GG, female *Diaptomus* antenna

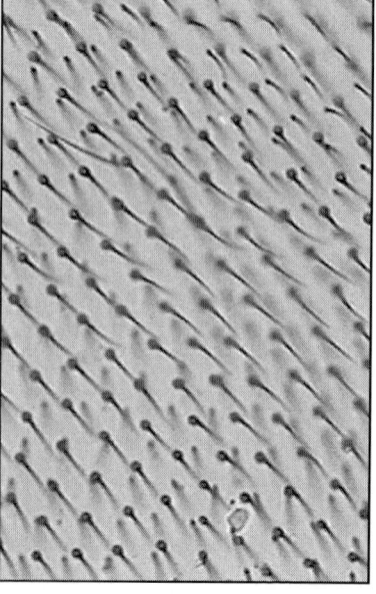

GG, fly wing-2

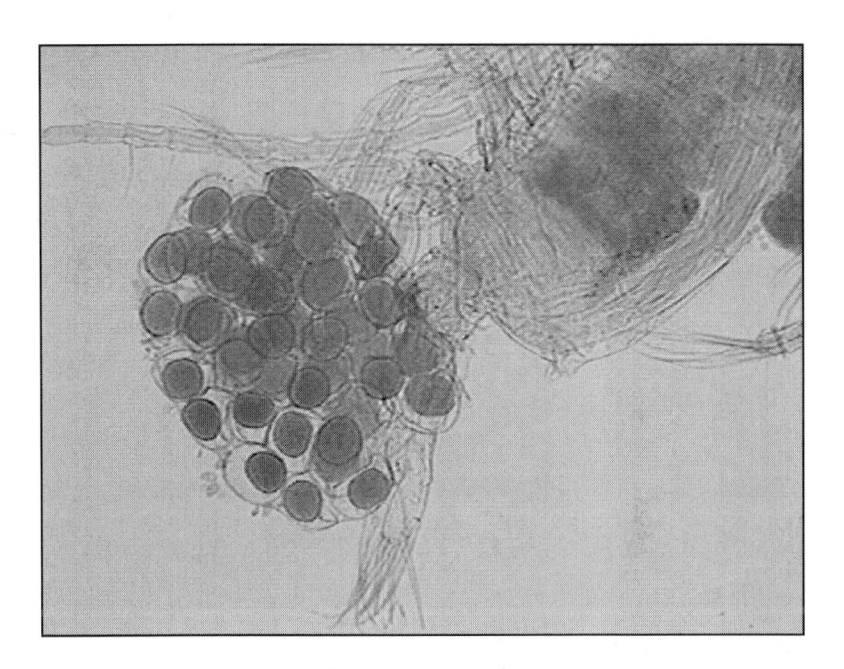

GG, *Diaptomus* ovisac

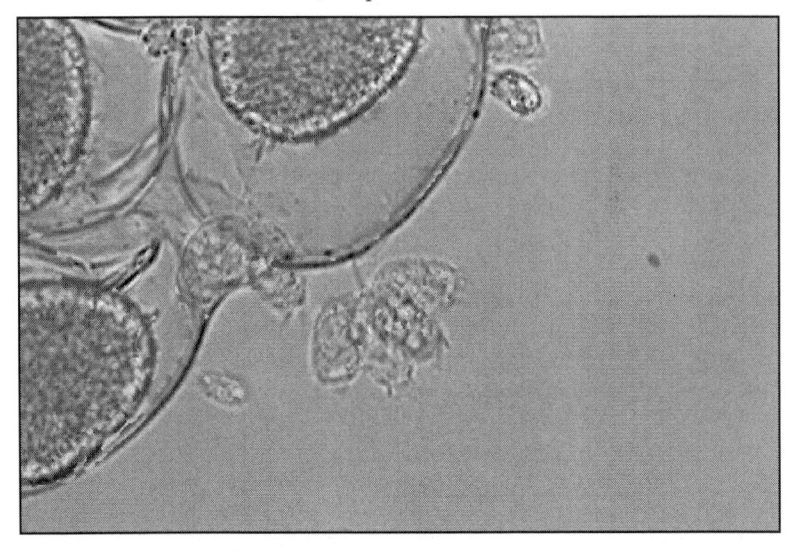

GG, epizoics on Vorticella

The interesting things to note in these examples are the epizoic peritrichs (protozoa in the group of the vorticella) living on the egg sac. They were perfectly fixed by the lactocupric fixative, you can even see the cilia in the open ends of the cells. They were fixed and kept in the fixative for more than a month.

GLG.- Glycerinated lactic gum

lactic acid............................... 25 ml
gum arabic............................ 40 ml
glycerin……………….………...20 ml
Karo*……………..…………...…10 ml
  *or Larry's fructose syrup

The above is my recommendation as a substitute for the Hoyer, Berlesse, de Fauré and other chloral hydrate based formulae. It is recommended for botanical sections cut with a microtome, including stained ones, or for arthropods or other subjects cleared in lacto-glycerol. Of course this and the PVA-L are the appropriate mediums for chitinized arthropods cleaned of their soft parts with potassium hydroxide or sodium hypochlorite. Small and soft subjects can be over cleared. Try the lactic acid methylene with blue staining (one that even I have not tried).

Examples

GLG, head of mosquito larva (below).

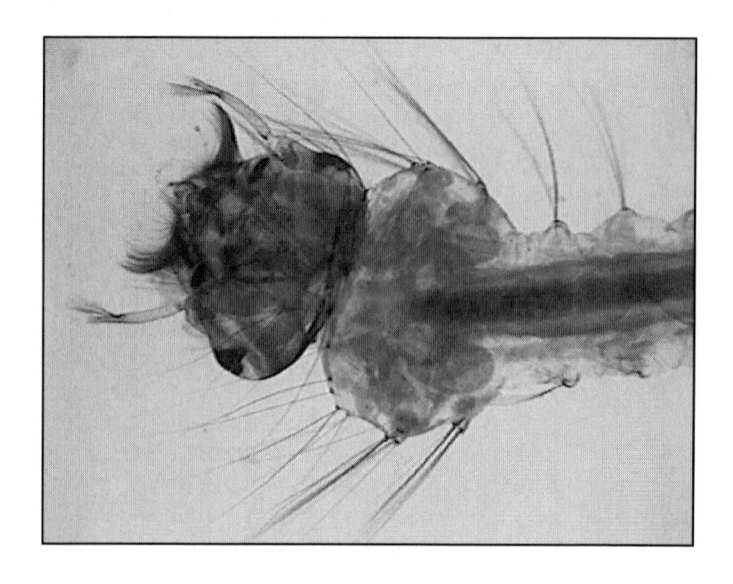

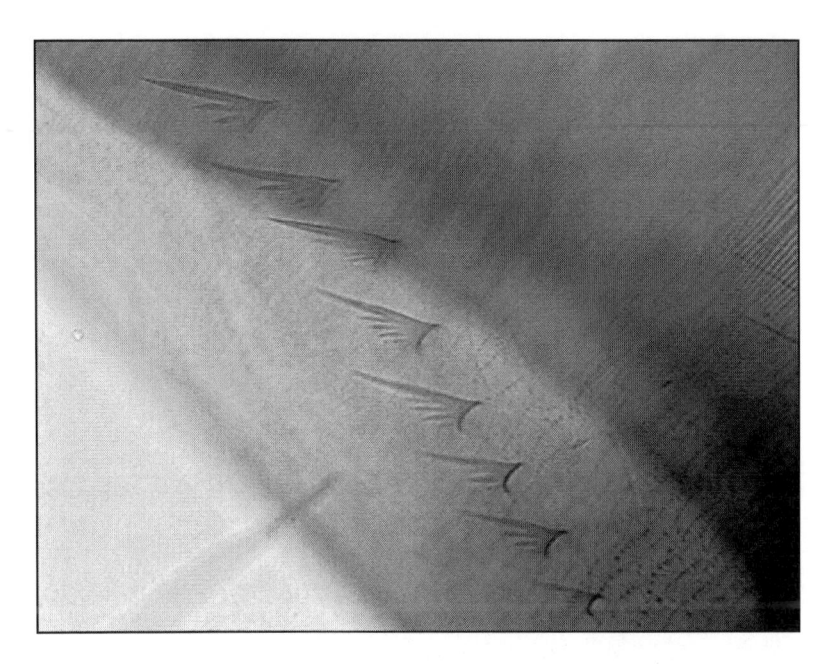

GLG, pecten in airtube of mosquito larva (above).

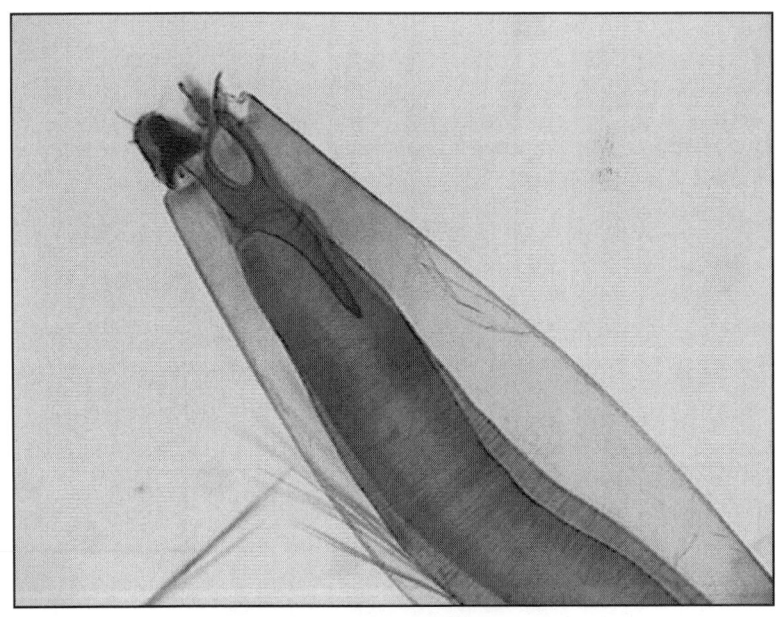

GLG, airtube of mosquito larva

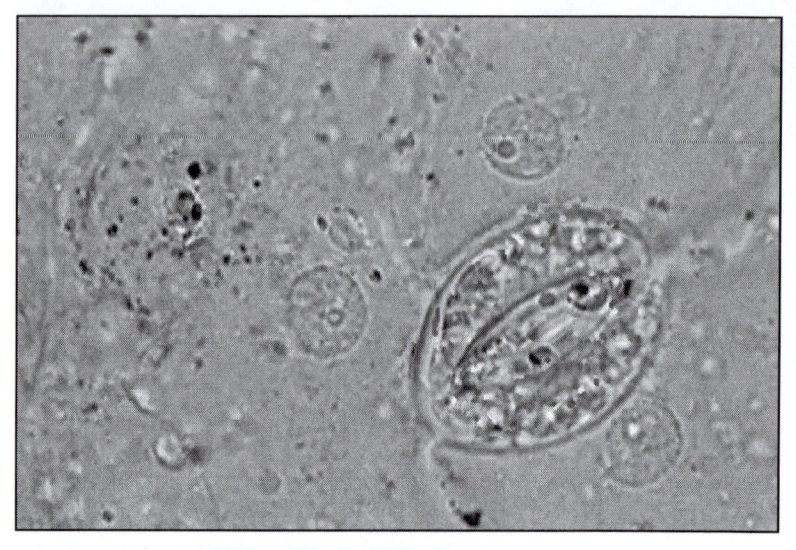

GLG, epithelial cell of leaf underside

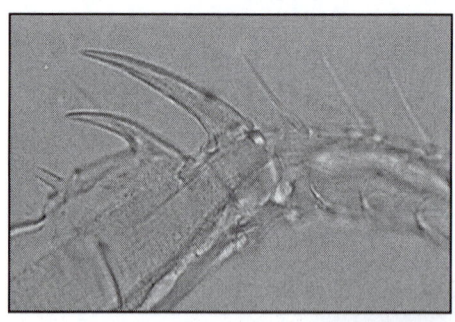

Left. GLG, *Diaptomus* male antenna

In my bottle this formula is highly fluid, and I first suspected it was of little use. But it is really an easy to use media. Drying time depends on the climate I think; it took much more time here in moist weather. If you mount directly some more or less opaque objects, they clear slowly in 8 to 12 hours. Muscles can be made invisible in the medium term. Generally this is the effect that users of this type of formula are searching for. Additionally you can clear your subjects, before mounting, with lactoglycerol, or lactic acid, if they are

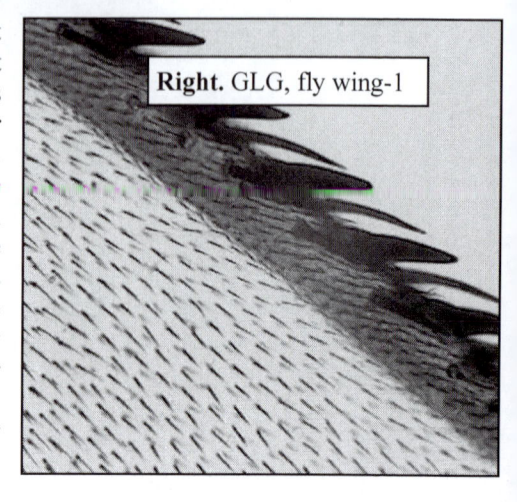

Right. GLG, fly wing-1

really dark.

The GLG gives results very similar to the PVA-L. Probably the degree of clearing imparted by each formula could be regulated by changing the quantity of lactic acid included. It dries firm in some hours, but it is a good idea to seal the cover slips.

Mounting in gum arabic media

With an appropriate tool pass the objects from water or an aqueous medium to the slide. With an absorbent paper eliminate all the water you can. Add a very tiny drop of medium, position your subjects as desired with the aid of two needles. Lower over them a carefully estimated drop of the gum media. It is better if it is a little undersized. Apply with care the cover slip. When you apply the weight the gum must reach the borders. You can hurry up the drying process by applying heat with an electric bulb, or with the microwave oven, as explained before.

If you live in a climate with 30% Relative Humidity or less this is all you need. But over 40, 50% RH you need to seal the coverslip.

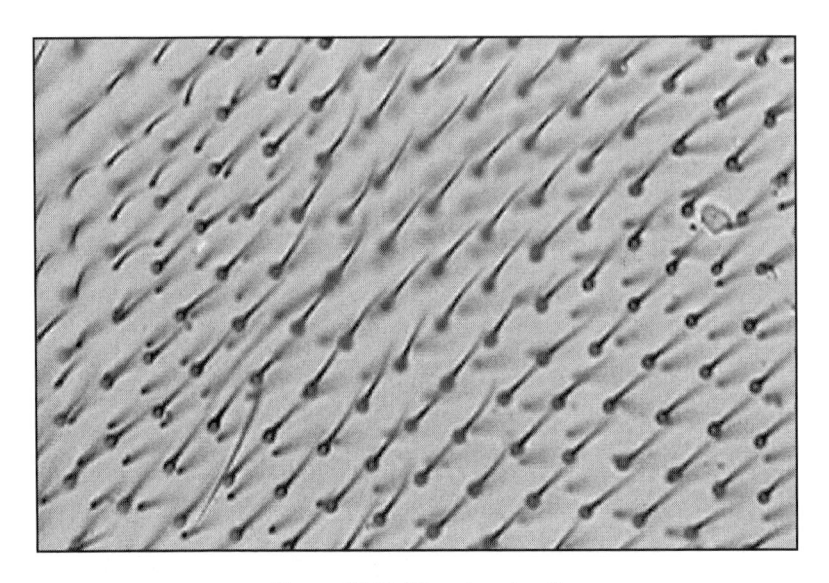

Above GLG, Fly wing detail.

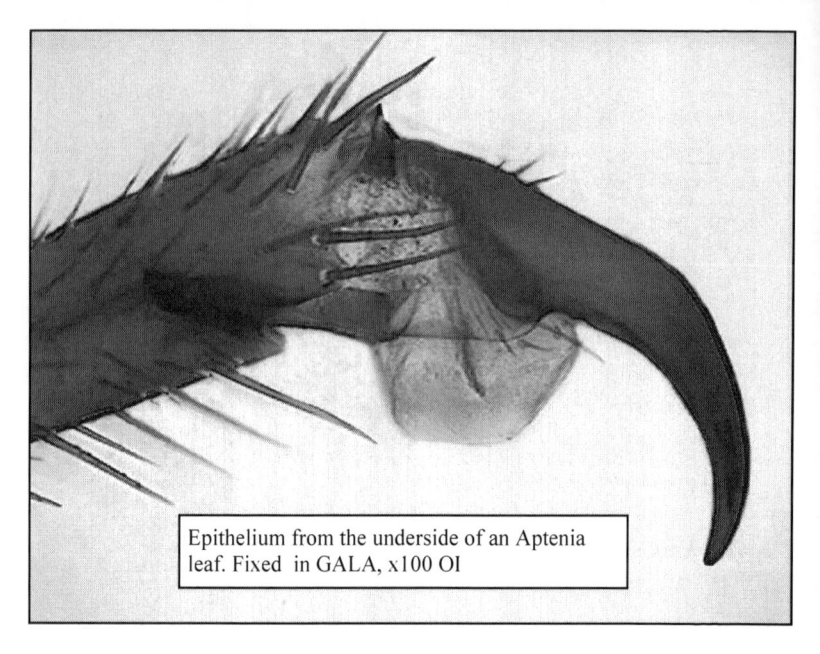

Epithelium from the underside of an Aptenia leaf. Fixed in GALA, x100 OI

MOUNTING MICROSCOPIC SUBJECTS
Chapter 4 - The Glycerin Jellies

All the pictures were taken with a National Optical microscope, equipped with an integrated digital camera and software, to control the picture taking parameters, from slides mounted in Kaiser Glycerin Jelly. The pollen pictures were assembled with CombineZ. Illumination was provided through a modified COL filter that will be soon described in a Topical Tip (see the original articles by Paul James).

The title picture is the tip of the tarsus of a leg from the cockroach *Periplaneta americana*. It has two claws. The second claw is beneath the visible one. The yellow feature between both claws is an adhesive organ that allows the roach to climb vertical surfaces easily.

Gelatin is an industrial product derived from collagen that is present in most animal organs. It is most abundant in bone and skin, which use to be discarded when animal bodies are processed for food. A treatment with hot water recovers the soluble portion in the form of gelatin that is dried and sold as thin plates, or powder.

Gelatin has the useful property of forming a jelly when it is treated with hot water. And as reagents are readily mixed with glycerin, it provides the perfect solution to the problem of solidifying glycerol (see the 'glycerin' section in the first part of this series).

Please! Do not try to use gelatin jellies without an antiseptic. If you do, in one or two weeks you can be the proud owners of an assorted collection of fungi and bacteria. (See the article on Gum Arabic media for a discussion of some antiseptics and a justification for my selection of Listerine.)

GJ - Kaiser's Glycerin jelly.-

The classical formulation is that of Kaiser (1880) which is also the best to be used in my latitudes. Many authors assign to it a RI of 1.47.

To make Kaiser's Glycerin Jelly we (and most of the professionals also) can use unflavoured Knox powdered gelatin, from the supermarket, that comes in a box with four envelopes of 7g each. So I adapted the formulas to allow preparations with an envelope of powdered gelatin which provides 7g.

water -   21g
glycerin - 12g
gelatin -   7g
listerine - 2g

Sprinkle the gelatin over the water which has been already well mixed with the glycerin, and leave it to soak for at least five minutes. Melt it over a low heat, or better in a 'double boiler'. Gelatin melts around 40°C. Add the antiseptic away from the heat and stir very slowly to avoid air bubbles. But don't be alarmed if thousands of the tiniest bubbles develop in the liquid, it is normal. Leave the flask in warm water for a while and in a few minutes the bubbles disappear. Pour the medium into a wide mouthed bottle. This media keeps very well. To mount your subjects follow the instructions below.

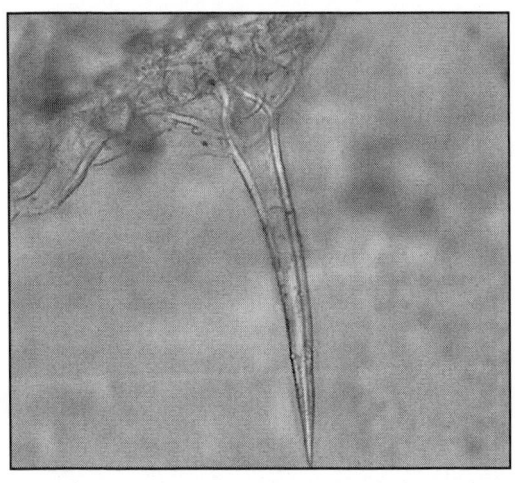

A "hair" from the epidermis of *Pelargonium*

FJ. Fructo-jelly.

>fructose (Karo) - 30 ml
>water - 10 ml
>gelatin - 7 g
>glycerin - 6 ml
>antiseptic - 4 ml
>sodium chloride - 1 teaspoon.

Mix Karo and water with the glycerin. Dissolve in the table salt. Sprinkle with the gelatin and leave it for 5 minutes for it to swell completely. Melt over a low heat or in a double boiler.

The RI must be around 1.45, with a pH of 6.6. This is the other formula that gives me more stability in the high summer temperatures. It is less solid than the Kaiser formula.

Examples

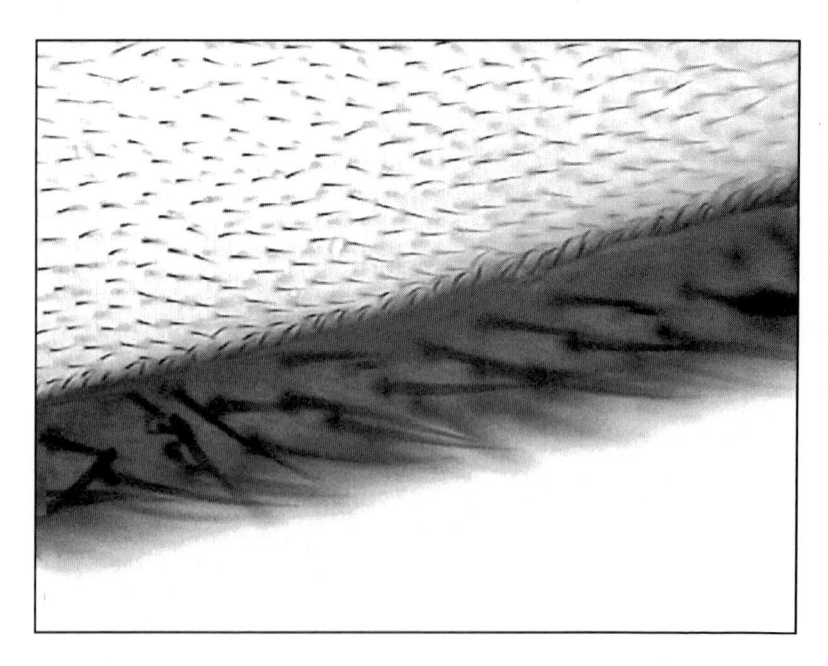

Wing border from a diptere of the family Tipulidae. Objective x40. Opposite page, top. Air bubbles. This is the more common and very frequent problem with Glycerin jellies.

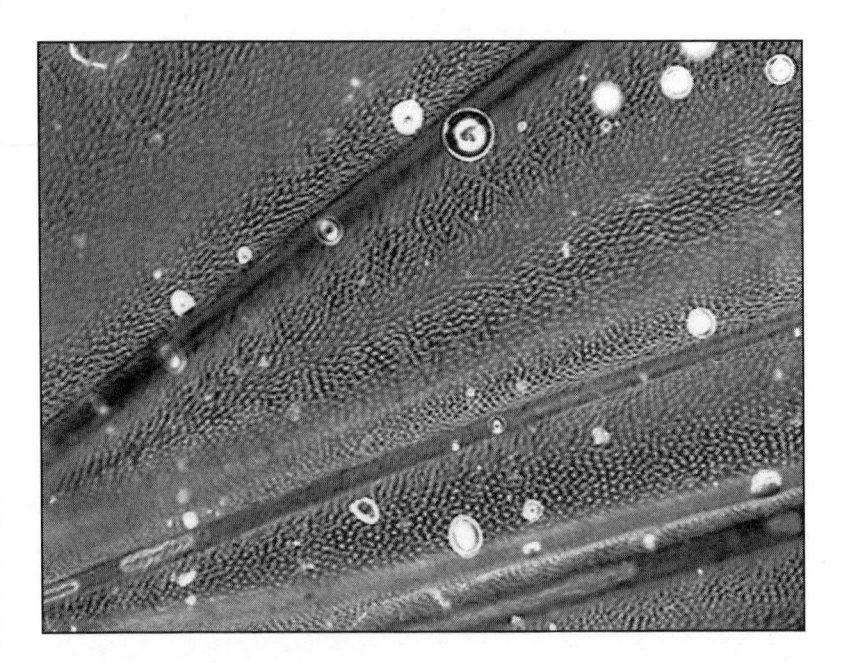

CGJ - Chromed glycerin jelly
Regarding the RI of this formula, see my comments in the introduction to part two. Reportedly it is a jelly formula that is still solid above 60°C, which could make it very useful in tropical climates. I give here the original formula adapted to one gelatin powder envelope.

>	Water - 120 ml
>	Glycerin - 25 ml
>	Gelatin - 7 g
>	Listerine - 10 ml

Soak and dissolve the gelatin in half the water. Add glycerin. Warm the remaining water and use it to dissolve the alum. Mix and add Listerine.

For those that have some interest in the chemistry of this formula, chrome alum is a double sulphate of potassium and chromium. The common alum is aluminium alum, a double sulphate of potassium and aluminium sold as a translucent 'stone' that in times of razors, men used to stop bleeding of little cuts on their face. The alum of chrome is most used in tanneries for leather-dressing, but has multiple industrial uses, and is said to be easily obtained, which is not my experience.

To date I've been unable to prepare this formula, my supplier sent me the wrong alum, and I couldn't obtain the small quantities of reagents I need to prepare my own chrome alum. I publish the formula expecting that someone could be more fortunate than I've been.

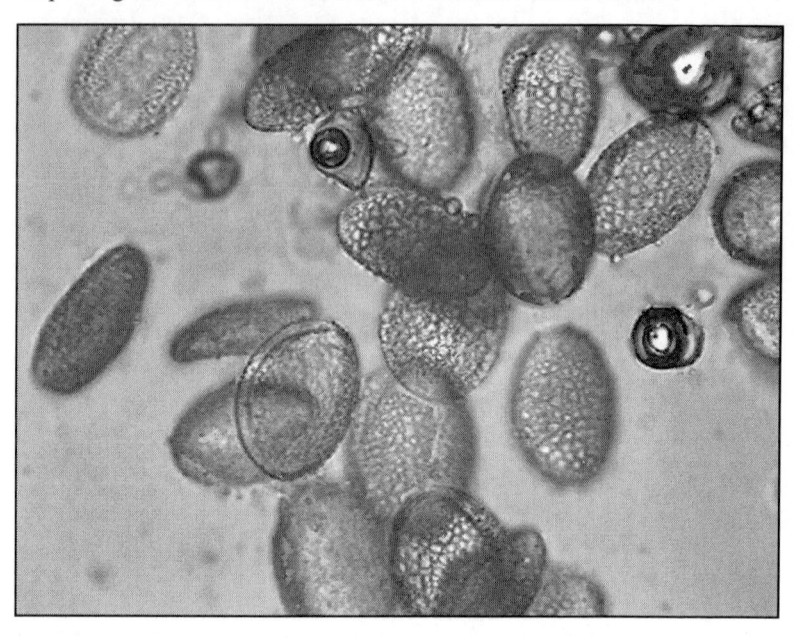

Above & below. Pollen from a flower of the family Liliacea.

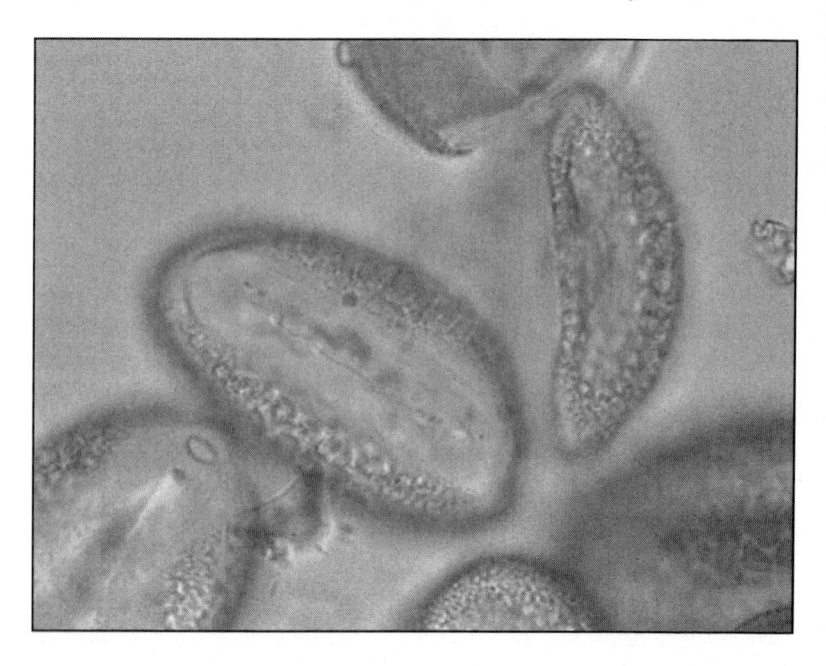

BJ - Borax Glycerin Jelly

This is a formula proposed by Fisher in 1912. Borax is sodium borate. My sodium borate is a chalk white granulated powder. Completely dissolve the borax in warm water. Add glycerin. Soak the gelatin and melt.

> Water - 55 ml
> Borax - 1.5 tsp
> Glycerin - 5 ml
> Gelatin - 7 g
> Listerine - 5 ml

It is said to be liquid at room temperature, but my preparation turned solid even with a double quantity of water and borax (sodium borate). At first I think the borax made the difference. But borax is only a mild antiseptic and an alkali, both very useful properties for a mounting medium.

I think that I can solve the 'mystery of the solid borax jelly'. When I was a boy, my mother dressed (sized) the embroidered works she was so proud of with a gelatin bath. The dressed embroidery was pinned on a flat surface and left to dry. It remains well extended and flexible. But in those times gelatin was not Knox, and the animal gelatins were not well purified, they were mostly used in carpenter's works (and smelt very badly). My mother (in the 40's) sent me to the pharmacy to buy 'fish gelatin'.

Fisher published his formula in 1912. It is most probable that he has used fish gelatin. Fish gelatin solutions are liquid still to 10 or 12°C, when they gel. So you can use Fisher's formula in the modern solid way as I did, or try to find 'fish gelatin' as a thin solid plate or a powder. Tell me how it performs.

Anyway to date Fisher's Borax jelly is unusable here, because it doesn't solidify at the summer temperatures of Durango, and even with Listerine it developed a heavy cover of moulds.

Note for all jellies, and other water based media.

Being a water dispersed colloid, the jellies have a remarkable characteristic that they share with Fructose and PVA based media. Water soluble dyes can be added to these media to impart to them a medium heavy tint. When applied to the slides and thinned enough under the coverslip the tint is barely appreciated. But the dye migrates slowly to the mounted subjects, imparting them some colour. This is especially useful in the mounting of pollen for which the gelatines are the preferred mountants. Several dyes are recommended: Basic or Acid Fuchsine, Malachite Green, Cotton Blue (Aniline Blue), Safranin, etc. Applying dyes to the mountant media has another use. Some stained materials fade with time, bleeding dye from the subject to the media.

Mounting in the Jellies

J. Kiernan in his page on water based mountants remarks that anyone who could mount something in glycerin jelly, can easily use any other mountant. Here are some of the tricks people use to do the job.

1) Howard Webb mounts his 'daphnia' alive in a drop of Glycerin Jelly he melts with a cigarette lighter. So simple could be the technique. And a very similar method is recommended, by Jean-Marie Cavanihac but using a more friendly heat source: an incandescent lamp of 25W.

But you need to develop a lot of experience to melt the gelatin exactly at its melting point avoiding overheating, which boils the mountant, develops big air bubbles, 'cooks' and dries your subjects, and even can break the slide. It is not so easy, but you can learn of course. If others can do it, why don't you try?

2) In the February 1999 issue of Micscape Magazine you can find an article by Brian Adams describing a small heating plate to help in the mounting. It is ingenious but is only a step ahead of those methods discussed in 1).

3) And a little 'stove' to maintain the jelly melted is proposed by H. Zander, (March 4, 2002)."I have a dropper bottle of molten glycerin jelly sitting on a small hot plate near my microscope. The hot plate is one of those coffee warmers you can pick up in a flea market for a buck or two. It was too hot, so I bought a light bulb dimmer at a hardware store, and used the rheostat to cut down the temperature to "merely warm", which was good enough." It must be complemented by a small warmed table that helps in the further mounting steps.

4) In the old times the task was accomplished by a Malassez table. This was a metal strip (generally copper) of more or less 6 or 7 cm wide, twisted as an 'S' with right angles, and mounted on thin legs. One end of the strip was heated with a little Bunsen torch or an alcohol lamp. A temperature gradient developed through the metal ribbon, its intensity depending on the applied heat, and the room temperature that governs the heat loss. The gradient is stable enough to permit the selection of the temperature required to work with. The 'modern solution' is to use a heat source that is safer, but the Malassez table is easy to make and to use.

A heating table, Malassez style.

5) I solved the problem from my own tests with a stainless steel plate 50 cm x 6 cm mounted at each end over two wooden supports high

enough to allow a small alcohol lamp to be used to heat one end. The flame must be small, and you must watch carefully not to heat the metal too much.

I put my subjects on the slides, side by side with a little portion of jelly (more or less equivalent to a 3 or 4 mm cube) and I start my burner. After a very few seconds of heating I displaced slowly my slides from the cold end towards the hot one, until I saw that the gelatin started to melt, and prudently I go back again to prevent boiling. When all the jelly is evenly melted, I displace the slide to the colder end and apply the coverslip. With appropriate forceps I remove the slide from the heating table, put it on a horizontal surface and apply a weight. If you have to mount many slides in a row you need to stop heating the metal table from time to time to maintain an appropriate temperature.

6) And, what I think is the better solution is the home made oven presented by Jean Legrand in the MICROSCOPIES M@GAZIN. It is simple to build and operate. Works in the range of temperatures most microscopists need for their tasks, and has enough space to have at a uniform temperature the melted glycerin, the slides and cover slips, and the instruments needed for the manipulations.

My suggestion is that you don't try to use jelly formulas if you are not prepared to have the slides, cover slips and gelatin hot enough (> 40°C) all the time from start to finish of the preparation. The method to do that depends on the subject you want to mount.

Even if the gelatin media doesn't need a previous dehydration, and all them received the materials directly from water, for best results it is better to pass previously the subjects through some dilutions of, or, in some cases, even to pure glycerin.

The materials that are stored in alcohol must be previously passed to water or to glycerin, because gelatin doesn't mix with alcohol. Revise the methods for inclusion in glycerin given in the first part, especially the Seinhorst technique.

The mounted subjects clear in minutes to several hours. Cut the excess jelly from around the coverslip, clean with tepid water, cold water, and alcohol, and seal with nail polish; finish with a layer of automotive paint (or glyptal if you can get it). In the high summer, in many countries, the jelly can become somewhat soft. Be sure you file the slides horizontally. While the jelly is molten on the heating plate you can use the usual weights, if you wish. Or you can use an old useful tool.

Take a cylinder of aluminium 18 mm in diameter with a length of 30 mm. Make a hole in its side and screw in a large and thin cylindrical screw. Cut the head of the screw and fit to it a wooden handle. It now resembles a hammer. Make your slide (slides) and leave it (them) to cool, even if the jelly sets before it reaches the border of the coverslip.

With a suitable source of heat carefully warm your 'flattener' and apply the flat face of it to the coverslip. The gelatin melts and you can exert the pressure you need. Put the slide on a level surface, apply the usual weight if you want and leave it to cool.

Of course you can make some pretty slides if you have a hot plate (even so small as the Adam's proposal or as good as J. Legrand's oven) to heat your slide while you are working, if you have a very clean jelly and take the precautions and the time to exclude all the bubbles. When you take the slide out of the heat, the medium jellies in a hurry and your preparation is almost finished. It took probably less than half an hour from start to finish.

Other Commentaries: About the permanence of the jelly preparations I see one testimony of some slides in the British Museum of Natural History which are over 100 years old.

Sometimes air bubbles can be useful!! This one (below) is showing the good centering of the darkfield stop under the condenser.

The following two images are examples.

The brood pouch of a daphnia.

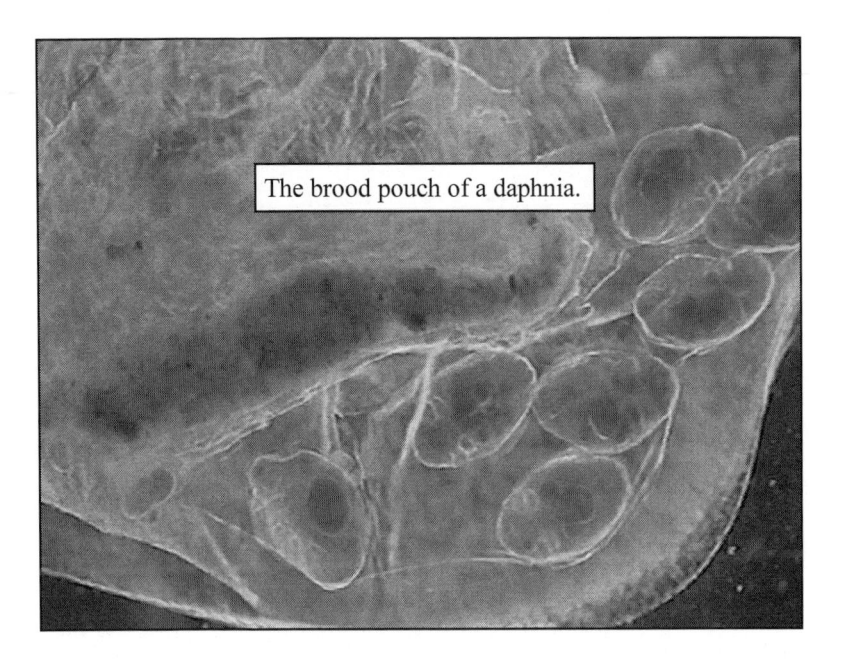

You can acknowledge how these specimens' processes are clear to see.

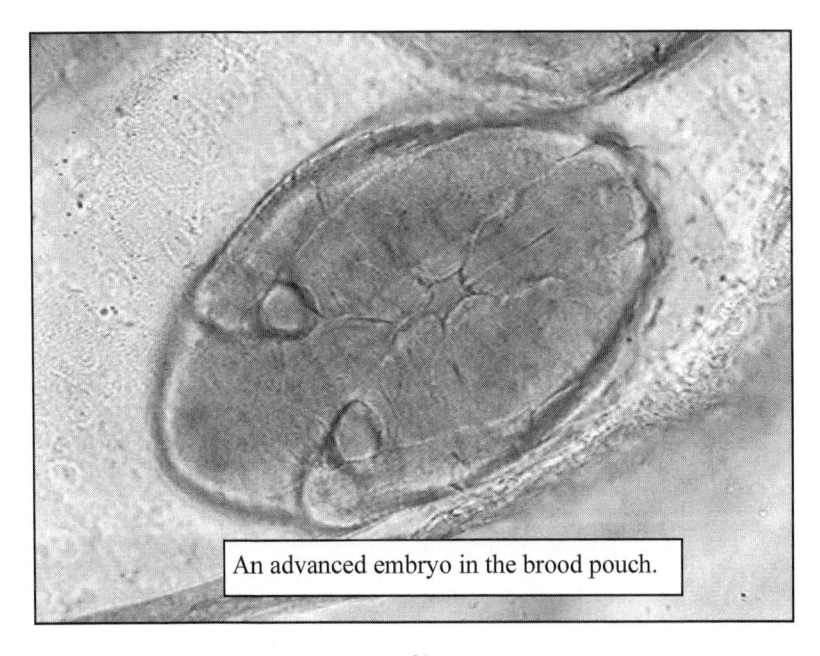

An advanced embryo in the brood pouch.

MOUNTING MICROSCOPIC SUBJECTS
Chapter 5 - Ten Years After.
(Written in 2010)

A rotifer from a shallow pond near Durango, Dgo., México, mounted in glycerin 10 years ago

JUSTIFICATION

Alexandre Dumas was the author of the famous book "The Three Musketeers". In his time of course they had no cinemas, nor movies, but they already used "sequels".

"The Three Musketeers" had its own sequel, which was called "Twenty Years After".**

**He even wrote a sequel for the sequel "The Viscount of Bragelone".

When I wrote the articles on the mounting media (2002 - 2003), especially the last one, I did not think at all of writing a sequel. No thought at the time, nor do I think now, to make a review of my preparations "Twenty Years Later."

However I believe that as it's now ten years since I started my first experiments, in early 1999, on healthy mounting media, safe for amateur use (or whoever wishes to try), is a good time to attempt a review.

SLIDES OR PICTURES?

Amateur microscopy has developed a lot since I found the first articles published on the Internet. It has increased the number of groups, and, especially, the used equipment offered has become decidedly at a quasi-professional level; with a profusion of famous microscope brands, and quantity of Phase Contrast or Difference Interference Contrast equipment. A researcher, a rotifer specialist, high-level, and with numerous scientific publications, told me that he does not have at his disposal the high quality equipment which the more advanced amateurs possess.

Along with the microscope, the essential equipment of any amateur at any level, photomicrography has developed, driven primarily by the availability of digital cameras (a dream no one dreamed of 30 years ago) at all levels of quality and price, many of them more or less easily coupled to a microscope.

The computer is now natural equipment for a student, and along with them the "webcams", currently essential auxiliary equipment, are purchased, which have now reached a resolution of 2 Mpx, with video capture speeds of 60 fps. Digital cameras have become common, with a quality at an unimaginable level 5 years ago, which now offer 12 and more Mpx resolutions. The range of prices and quality is so broad, and new developments so frequent, that it has become customary to consider disposable the "old" cameras. And with any of them, any microscopist, with a little experimentation, can access photomicrography.

Living organisms, their morphology, colour, and behavior can thus be well documented. And for an amateur this (now common, and previously impossible ability) generally meets all their needs. Therefore, the pictures, and video files, replaced to a large extent the slides, (difficult to prepare, maintain and store) that were the previous target of amateurs, as evidenced by the valuable review articles on old "slide-makers" regularly published at MICSCAPE, and such preparations are used to illustrate articles and discussions of microscopists.

However, even with the help of the frame-stacking programs to recover the field depth (e.g. CombineZ) it is impossible to capture in a single exposure all the details of an organism. In my work on the Bdelloid Rotifers I left a list of the details (and therefore images) needed by someone seriously interested in achieving the ability to determine the species. The more advanced amateur, seriously interested in the scientific discipline that he explores, will often want to re-examine some specimen, because of a desire to learn more about this species, or to use a newly acquired key, or for comparison with a new-found specimen, or to share their findings with a specialist.

Even if micrographs can be a goal in themselves, by their technical quality, or their beauty (there are forums to display them, and share them, and contests especially dedicated to promote them), the preparation of specimen slides, as a demonstration of technical skill, aesthetic satisfaction (see the incredible artistic images of old and new diatoms preparations) and also of scientific record, may also be a target for many microscopists.

Fig 1 and 2 (above) are two designs with diatoms. Dominique Prades pictures and preparations (see images at http://scopimages.free.fr).

If you wish to see some very old arrangements, visit also http://www.victorianmicroscopeslides.com/slides.htm

So I think that it's not trivial to show what happened to the slides I made 10 years ago to test the behaviour of different mounting media.

For those that want to try mounting something for the first time, I think that a careful reading of the Richard Howey article Permanent Slides: Pros and Cons is mandatory. There are many examples of subjects easy to manipulate and mount. Try them also with this different mounting media.

About permanent slides. Web link only...
http://www.microscopy-uk.org.uk/mag/artmar99/rhslide.html

The aim of the published series of articles was to eliminate health risks, and collect a series of formulae, useful over an ample range of different subjects, that use ingredients that were not banned or severely restricted.

Leaving aside the "wet mounts" in an antiseptic solution (usually the same sample water, with the added fixing agent) which, even sealed, can only be expected to survive a few weeks or a couple of months at most, of the 19 mounting media tested between 1999 and 2002, those below were discussed in the last article of the series, as candidates for intermediate or relatively long duration (5 or 10 years) which I think, could be the objective of an amateur microscopist.

Aqueous
GP – pure glycerin
GG – Glycerin Jelly
PVA-G – polyvinyl alcohol with glycerin
PVA-L – polyvinyl alcohol with lactic acid
Karo – or fructose syrup
GAF – Gum Arabic, Fructose syrup*

Resinous
GUM DAMAR
NPM – nail polish enamel
Norland 61
Additional
GLG – Gum Arabic, lactic acid
von Apáthy

A reading of the original articles is necessary for the understanding and use of this review. For each discussed media the address of the reference article is given under the title, go there and return to these notes.

Pictures included here don't claim contest quality, they are only added as witnesses of the formulae behavior after ten years in their boxes. If not otherwise stated, they were taken with the Logitech 9000. All were trimmed and/or reduced for to present them in an uniform size and shape.

PURE GLYCERIN (PG)
About chapter 2. Web link:
www.microscopy-uk.org.uk/mag/artdec02/wdmount2.html

This medium needs nothing more than the confirmation of its outstanding qualities. As I said, presenting it eight years ago, in museums all over the world there are all kinds of preparations mounted in glycerin for decades. The only contraindication is that it is a liquid medium, and hygroscopic, and therefore it is essential, but not easy even for professionals, to perform a very careful sealing of the preparations.

The range of subjects which can be mounted in Glycerin is extremely wide. It is interesting to note that professional researchers of nematodes and rotifers mounted them for study, and for the file of the types of new species, preferably in this medium. I found that preparations of periphyton and freshwater plankton mounted in 50-70% glycerin, have, 10 years later, the same quality, including almost the same colours they had when mounted. No dyes were used whatsoever. Preparations were ringed with nail polish. They were permanently kept horizontally.

Periphyton, washed and concentrated from aquatic vegetation of a shallow pond near Durango City, Dgo. México. Fixed with boiling water. Post-fixation with Lactocupric, washed with clear water 2 times, and glycerin added in approx. 5 stages, to reach 50%. Drops mounted, slightly compressed, and sealed with 1 layer of NPM, plus a thick another one, 1 week later.

Below. Fig. 3 - Chlorophyceae. 40x objective,

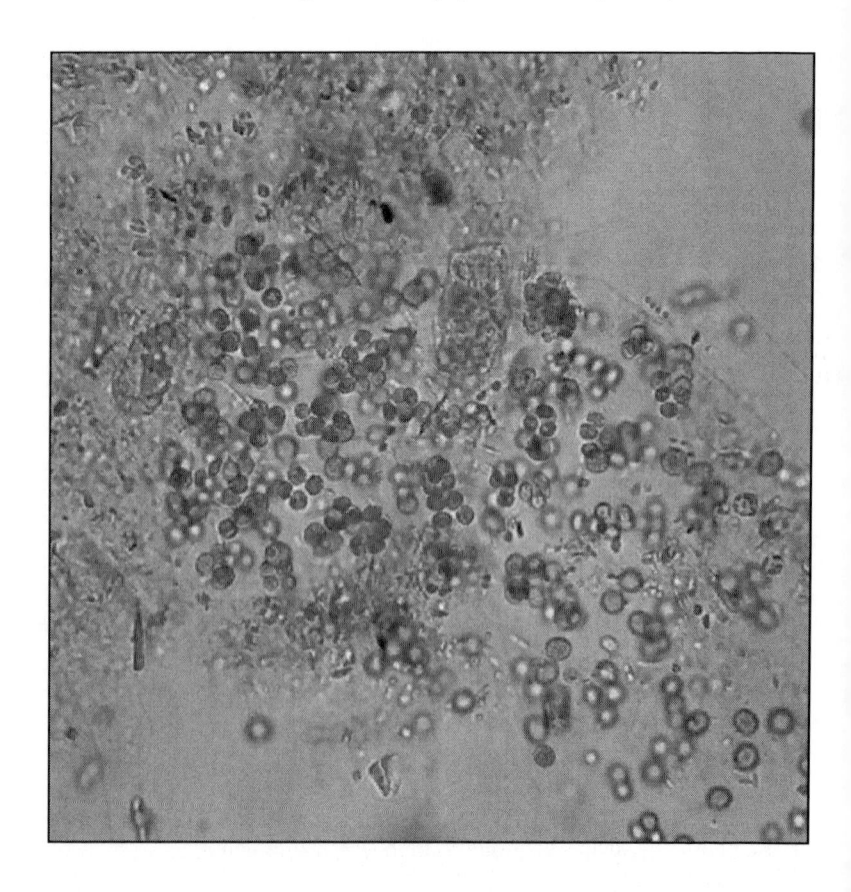

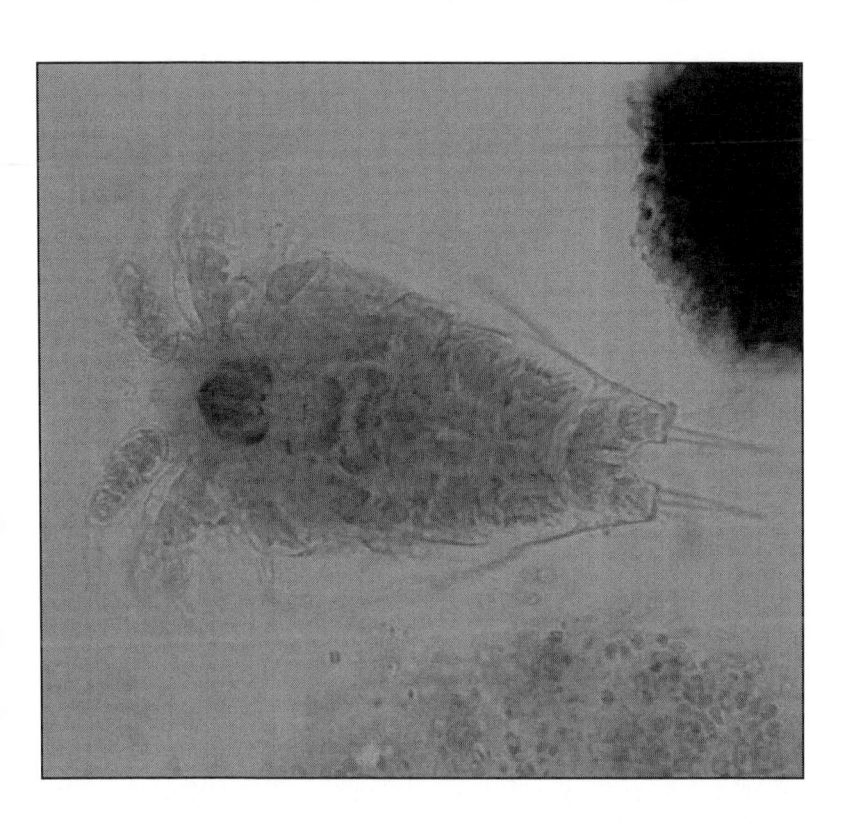

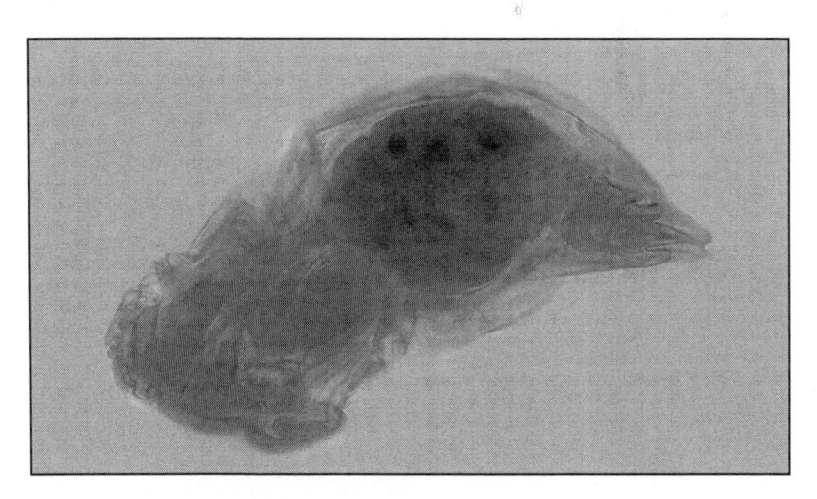

Above. Fig.4 - A copepod nauplius, *idem* rotifer
Fig. 5 – A Monogononta rotifer. 40x objective. Its opacity is just slightly more than when it was mounted.

GLYCERINE JELLY (GJ)
About chapter 4.
Web link: www.microscopy-uk.org.uk/mag/artaug03/wdpart4.html

This media is the answer (developed in the 19th century) to the problem of the liquid state of glycerin. The addition of gelatin allowed working more easily and safely, by providing a solid medium.

It is another mounting medium widely accepted, and, probably, the most recommended for amateur microscopists. And widely used by botanists, mycologists, phycologists, entomologists, and histologists. The best known formula is undoubtedly the one of Kaiser (1880), which is extremely useful even in the tropics, if preparations are kept in the dark, cool, and ventilated. The ingredients are very easy to get and combine. In addition to mine, there are on the Internet many instructions to prepare, and to use it.

Interestingly, this formula has been a means of identifying that MICSCAPE is being used not only by amateur microscopists, but by professionals also. The formula given by me in 2003 was published in 2006 by the Univ. of Minnesota (without declaring the source of it) in their formulary. (http://bipl.umn.edu/files/Mount.pdf) The origin is easily identifiable. I modified the formula of Kaiser in order to adapt it to the gelatin source at my disposal (Knox powdered gelatin) that is sold in bags of 7 g, and to the humidity and temperature in my lab in Durango, Mexico. In addition to that, until I published my formulas, the disinfectant that was incorporated in Kaiser was phenol (now banned for amateurs). I changed it to LISTERINE. And the preparation instructions provided are practically the same as I gave. It is gratifying that MICSCAPE, while not being a "magazine controlled by reviewers", can be a source of useful knowledge at the professional level. But not always easy to verify the transfer.

Of course Glycerin Jelly is used molten, and one of the most elegant media to work with without trouble, as you can work with any other liquid medium, and having long enough time even for a micro-dissection, without risking adding more air bubbles that is unavoidable, and using the efficient heater of J M Cavanihac.

Also for gelatins, the suitable range of subjects is enormous. It should be remembered that it is a good precaution to embed the subject in glycerin, before mounting it. There is no need to take it to pure glycerin, a concentration of 50% is satisfactory.

(Above) Figure 6 - Wing of a crane fly, as the borders were not sealed, and no spacers were used, the compression of the coverslip as the medium dehydrates slowly, caused some cracks in the chitin (below).

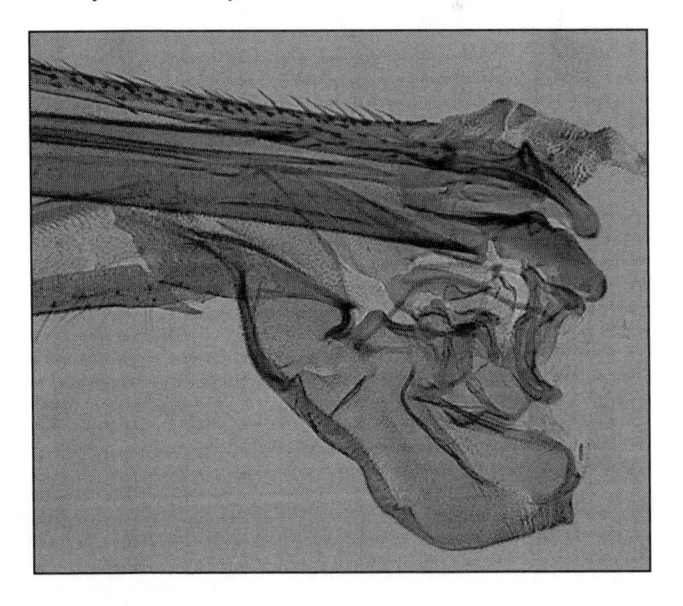

Of course there are many published recipes using glycerin jelly, which usually only differ in small percentages of the ingredients, that are not even worth trying, as the antiseptic used to keep it free of pollutants are generally all banned or discouraged today for health reasons. You can use any antiseptic compound soluble in water and miscible with the jelly. Clear solutions of benzalkonium chloride are also acceptable.

PVA-G (Polyvinyl-alcohol with glycerin)
About chapter 3b.
Web link:
www.microscopy-uk.org.uk/mag/artapr03/wdpart3b.html

This media is not so easy to prepare outside the U.S.A. Although I get a good consistency PVA syrup in Mexico, is not as clean as the ITOYA O'Glue I recommended in the original publication, which I am still using now. From declarations of South American and European correspondents it seems that it is difficult for them to identify a reliable source for this product. Except of course for those who have relationships with a well equipped professional laboratory, the chemical powder of good quality is extremely difficult to buy, and expensive also.

It is a pity, because the optical properties and conservation are very good, dries quickly and stays clear. I still recommend to strive to find a ready made water clear transparent adhesive, with the consistency of honey or syrup, which most probably is based on PVA (even if the label doesn't declare this) by making a tour of the stores that provide materials for schools, or offices, or crafts, or shops selling holiday ornaments and accessories. I find now that it is also appropriate to seal the preparations because some show some shrinkage of the medium around the edges of the covers. It is logical, because the PVA is in aqueous solution. The ones sealed have not had that problem.

Howard Webb (see his articles in MICSCAPE) has used this medium to mount his cladocerans. At my request he kindly informed me about the behaviour of the medium in these terms:

"I am still using PVA-G (just mounted some more daphnia this evening). Looking back at old slides, they seem to have held up well in general. Most of the shrinkage occurred in the first month (no prep and too much water), and there is no noticeable deterioration of the specimens. I have not sealed them in any way (as I did with jellied glycerin).

The one thing I have noticed is a lot of condensation on the slides. I have them stored in plastic slide boxes, and it looks like there has been some evaporation and condensation of something volatile. Almost all of the slides have a mist of something on them (glycerin?). They clean up fairly easily, but definitely need a cleaning."

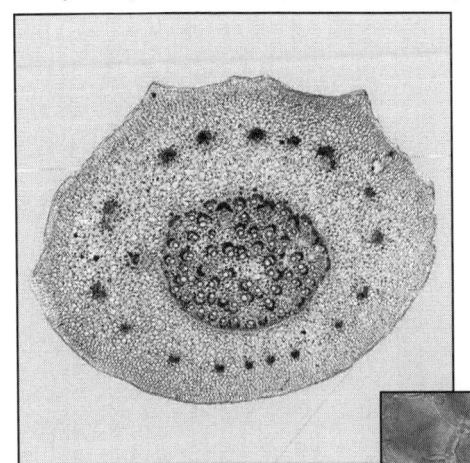

Left. Cross section of a stem of *Epipremnum aureum*, prepared with the mesotome. The image is a mosaic of 4 pictures, objective 4x.

Below. A package of vessels in the "stele" of the same. 40x Objective.

Below. a bunch of spines (probably sensory) located laterally in body segment of a mosquito larva.

All PVA-G Mounted.

Right. Topographical picture of the caudal end of larva.

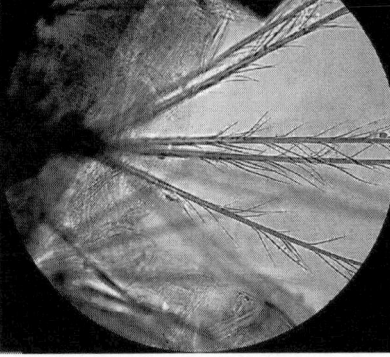

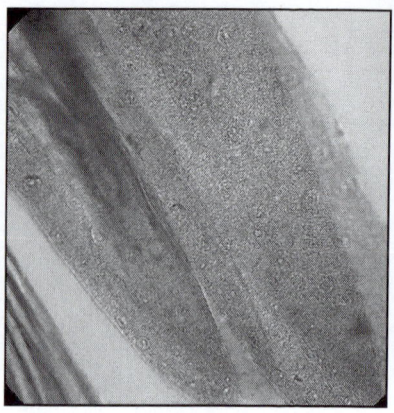

Gills of a culicine mosquito, objectives 10x.

They still show the nuclei of cells that form the branchial leaves. Taken with Logitech zoomed to 2.5 Mpx, more or less.

The last four images are from a thick preparation, with coverslip supported by 1 mm thick supports. Larva fixed for 48 hours in 70% alcohol, passed through 20% glycerol for two hours before mounting. Sealed with NPM.

Note: Unfortunately I lost the preparation of *Aptenia* epithelium mounted in PVA-G. I think it would have performed better than the PVA-L mounted, illustrated below.

PVA-L (Polyvinyl-alcohol with lactic acid)
About Chapter 3b.
Web Link:
www.microscopy-uk.org.uk/mag/artapr03/wdpart3b.html

There is only one thrip preparation, photographed and published years ago on a dark background. The preparation is firm and dry, without staining of any kind, with very good transparency and cleanliness. It was not sealed. That led to more than desired evaporation of the solvent, and the coverslip compressed the insect which deformed the glass, causing a small surface crack! But there is no shrinkage at the edges. It is, in all aspects, similar in quality to Gum Damar. It is worthwhile continuing to try this media, much easier to prepare than the Damar, but remember to seal to avoid this problem. I used two photos to make a collage.

Pictures on next page.
Fig. 14 Thrip, whole, obj. 4x. Fig 15 Thrip, antennae obj. 40x

Aptenia epithelium was mounted without colouration, directly after being fixed with lacto-cupric. The epidermal peel is very well preserved, but is practically unusable (and it is overall less than photogenic) because it lost the colour of the chloroplasts, and the medium has an RI excessive for this transparent subject. Only a condenser central stop of 15 mm in

| Fig. 14 | Fig. 15 |
| Fig. 16 | Fig. 17 |

diameter, creating a Circular Oblique Lighting, lets me view the stomata using low light. But this preparation shows that the PVA-L can keep perfectly vegetable sections. It should be checked whether a potential subject accepts and retains coloured preparations, but its acidity is high, which could attack the basic dyes.

Fig.16, Fig 17

Aptenia sp. epithelium. Fixed in Lactocupric. Mounted in PVA-L. Image taken using a Canon Powershot A75 (3.2 Mpx) handheld over the objective. 100x Obj. Illumination dimmed to a minimum compatible with picture taking. Chlorophyll is totally bleached, but histology is preserved. Fig 15 is the reduction of the original 3.2 Mpx picture, Fig 16 is a crop of the nucleus from the original.

As in other preparations this isn't sealed. The media is completely clear, transparent and reduced to a thinnest sheet. But there is no retraction on the edges.

Not all preparations were successful.

A mayfly larva was mounted in PVA-L and at the time it allowed great pictures in Dark Field and Rheinberg. It is now completely broken down and surrounded by an elliptical gas bubble that was rejecting the mounting medium around the larva, carving irregular channels and islands of varied relief. My hypothesis is that the larva, although I have recorded that it was fixed in Alcohol 70 was not fixed (Oho! That's shameful!)... or poorly fixed. I think the generated decomposition gases slowly created the bubble. This means that failure is a fault of the technique, and not of the mounting medium. No retraction at the borders.

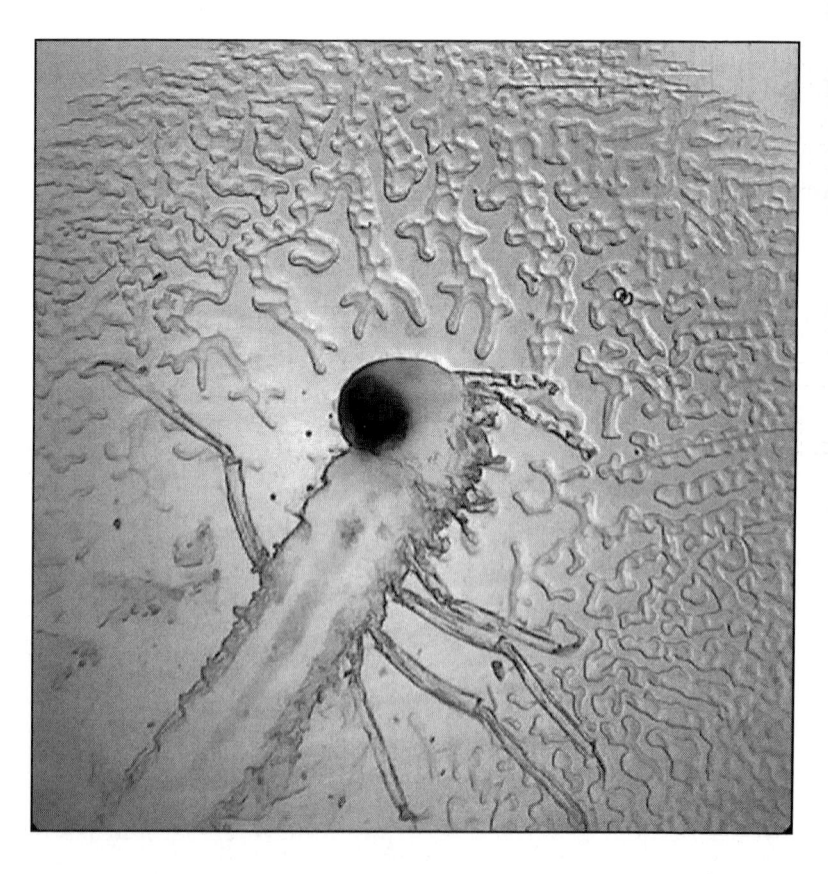

Above. ephemeroptera mouth parts—general view
Opposite page. Mouthparts of the mayfly larvae.

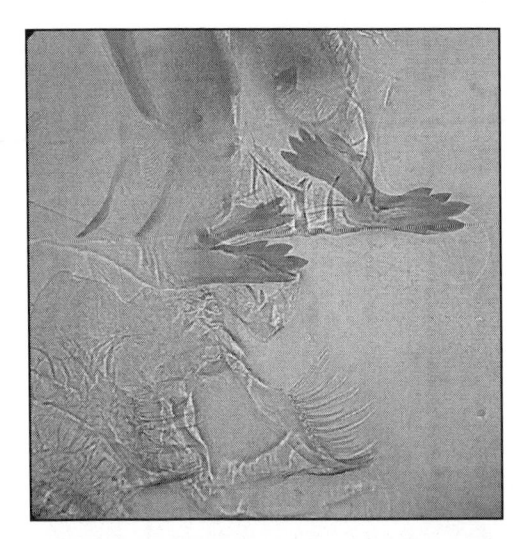

Below. Tangential cut from a geranium stem. Mounted to observe the spiral vessels, and calcium oxalate druses. Fixed in AFA, without colouration. The vessels look good, as well as crystals. The medium is clear, transparent, dried too much and the coverslip is deformed, as in the case of the thrip. The photo was taken with COL (15mm stop) There is also no retraction.

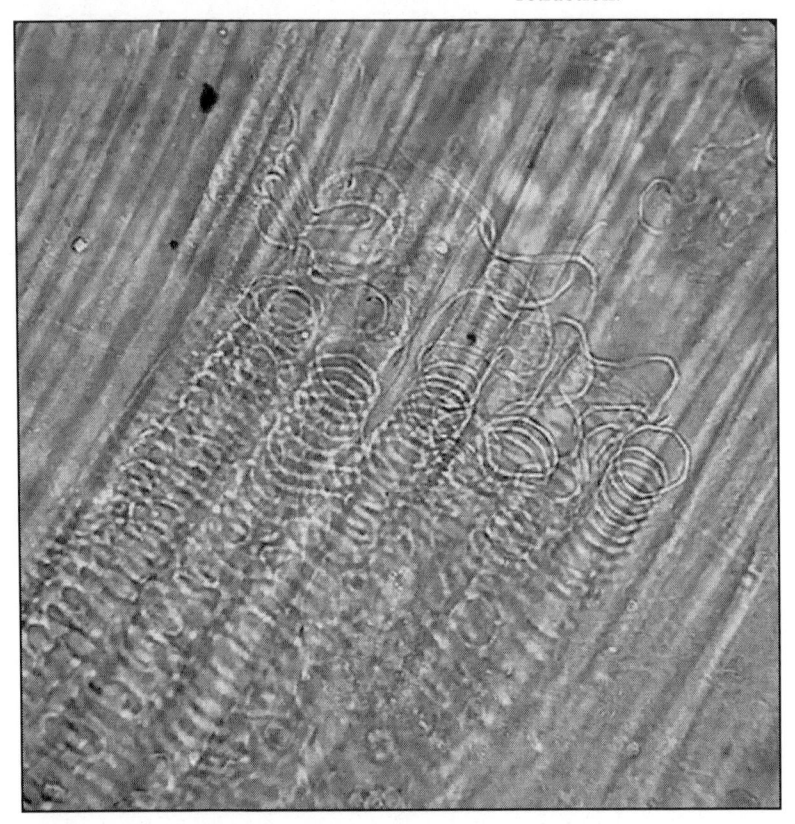

FRUCTOSE SYRUP and KARO
About Chapter 3a
Web Link:
www.microscopy-uk.org.uk/mag/artjan03/wdmount3.html

It is an inexpensive and extremely useful medium. Although does not seem to be so common in Europe (and I do not know what is available in the rest of the world) KARO is a common product on the shelves of supermarkets in the United States and Latin America.

Where it doesn't exist, crystalline, or powdered Fructose, is surely available in the dietary sweeteners section of the supermarkets. With this it's easy to prepare the syrup of Larry Legg, which is practically equivalent:

http://www.microscopy-uk.org.uk/larry/sugar4.html

Fructose Syrup is thus an absolutely safe mounting medium, easy to prepare and to use, and available to a very large audience. It is widely used by mycologists and phycologists as a preferred mounting medium.

The chloroplasts of *Aptenia* leaves epithelium, fixed with lactocupric, retained their green colour. Karo also retained the Gentian Violet I applied to onion epithelia. And copepods mounts are among the best I have.

The image of a copepod, used to illustrate Rheinberg illumination in the 3rd part of the articles on the Logitech webcam, is from a Karo mounted female.

I think that its utility, availability and even a good refractive index (1.48) recommends Fructose Syrups, handmade or commercial, as a general mounting media for amateurs. The only problem, not a minor problem for novices, is that it exerts a heavy osmotic pressure, similar to glycerin. And for delicate materials also needs a step by step mounting protocol.

Right.
Fig. 21 *Aptenia* epithelium, a stoma with still coloured chloroplasts (100x – Lactocupric Fixative. – Karo.

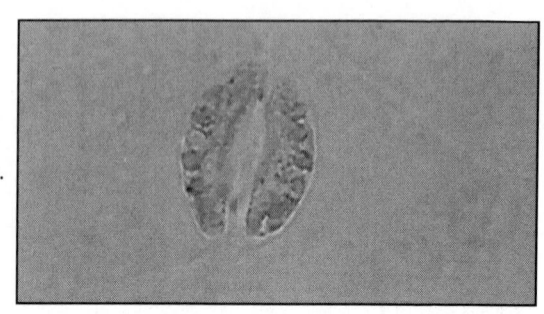

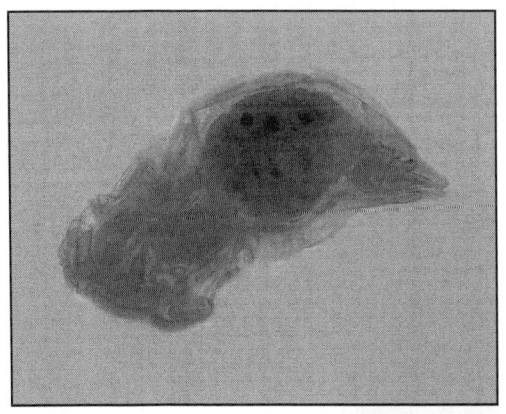

Left. Fig. 22 Muscles that move the swimming legs coxae of a copepod – Lactocupric Fixative – Karo.

Below. Fig. 23 Onion, epithelial cells (10x) stained with gentian violet.

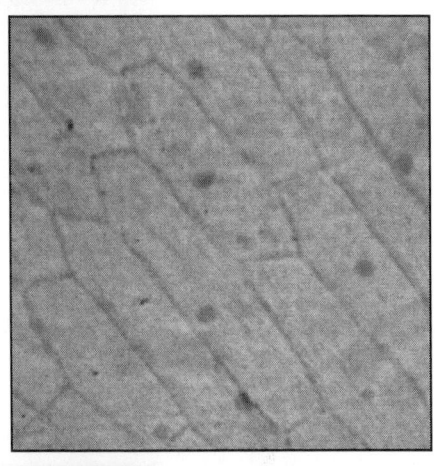

Focus on surface. Karo – Fixed with AFA.

Below. Fig. 24 Nuclei of onion epithelial cells, coloured with gentian violet (100x). Karo – Fixed with AFA.

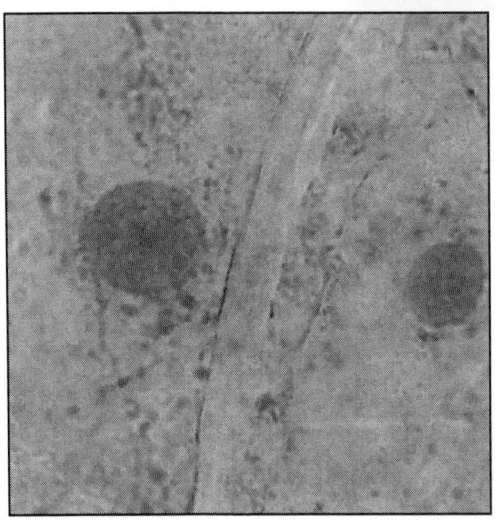

This ends this small work on alternate and safe ways of preparing and mounting specimens for microscopic study without using chemicals now deemed unsafe for enthusiasts to use.

I hope it proves helpful for existing microscopists and future generations.

Walter.

Finale
Editor's Note:

The 'Finale' section below was originally presented online in 2003, i.e. seven years before the 'Ten Years After' in 2010. Some of the text in the 'Finale' reflects this, but overall both the online and book editor felt that this was a suitable ending for a book form of the suite.

If you count them you must arrive at 20 safe mounting media discussed in this series. They provide mounting media for all the materials you may wish to prepare. You don't need, even for a medium term of permanence (5, 10, 20 years?) to have resort to the expensive Canada balsam. And remember that, even if this mountant cannot be excessively dangerous because of the few occasions you may use it, and the small quantities involved in any mounting operation, the real danger is from the solvents (all of them toxic, and flammable) which you must store and use when required.

Leave the Canada balsam for the professional taxonomists that legitimately search for 'HYP' (Hundred Years Permanence) for a type species, and have lots of fun time making your collection of really professional looking slides.

List of experimental mounting media

>AW, antiseptic water: sample water with added fixative
>PG, pure glycerin
>FMM, fructose mounting media (Karo, or Larry's fructose)
>CPG-pva, Commercial Pure Glue, based on PVA
>AG, alcoholic glycerol
>FG, Fructose glycerol medium
>FGL, lactic fructose glycerol
>APT, von Apáthy Gum Syrup
>GAF, Gum Arabic Fructose.
>GG, Glycerinated Gum
>GLG, Glycerinated Lactic Gum
>PVA-L, PVA formula with lactic acid for high clearing action
>PVA-G, PVA formula with glycerin, for a mild clearing action and less acidic medium
>GJ, Kaiser's Glycerin Jelly
>BJ, Borax Jelly (Fisher's type glycerin jelly)
>FJ, Fructo-jelly
>CGJ, chromed glycerin jelly
>NPM, Nail polish mountant
>Norland 61, a synthetic commercial glass-adhesive of high refractive index
>(DAMAR), gum Damar, diluted in Turpentine, or in Toluene
>Solidifiable media
>Liquid media
>Natural or synthetic resinous media

SELECTING A MOUNTING MEDIA

It is possible that only Canada balsam has the 'HYP' record, but if you are not a professional taxonomist trying to preserve type species for the eternity, who cares? And certainly the much cheaper Damar (diluted in Xylene or in Essence of Turpentine) can compete for the professional mountant title and (in turpentine) it is safer than balsam.

But I have given you 6 products that work directly from the container, (PG, KMM, NPM, CPGpva, AG, and Norland 61) and another 14 formulae for safe mountants, most of whose prepared slides can be filed for months, even years, some for decades. So which one of them must you use?

Almost any one of the revised media can do a good job as pictures in this article series have shown.

But nobody needs a complete set of mounting media. People are selective and each amateur has vocations and interests cantered on some groups of subjects.

This is my own discriminating analysis:

The most difficult to use.-

1) Two of them (AW and PG) although very difficult to use, are indispensable for all those that like to work with 'pond samples', invertebrates lab cultures, and similar materials. You must learn to work with them, especially to seal them in a reliable way. Pure Glycerin when VERY WELL SEALED is a professional quality media, filed in museums for decades.

2) In my opinion the other difficult ones are the group of 4 media of Gelatin Jellies. They have been handed down from folklore, coming from the 19th century, and certainly they are the most often proposed media for the amateur. But you need to melt them in an appropriate apparatus (most certainly you must make your apparatus) and even with some risk if you use gas or alcohol, open flames, matches or cigarette lighters. You frequently must filter them to have a really clear and clean medium, and fight against the ever present air bubbles under the coverslip, and they are thermally unstable, needing special care to store them and the slides made with them.

I must be convinced that my subjects cannot be mounted in other media, or that gelatin is the utmost media for some subject, for me to use it. As an example, mount a "daphnia" in Kaiser Jelly and in PVA-G, the lesser refractive jelly mount gives you a sense of depth that the PVA does not.

The less safe.- Gum Damar, although a relatively cheap and

efficient medium, with good optical qualities and a predictable life of about 100 years, is not a totally safe medium, and must be prepared and used by younger amateurs under adult supervision. It should be recommended only for grown ups, and very careful amateurs. But it really merits becoming a professional standard medium.

The specialized one.- The Norland 61 synthetic medium although having only a medium high refractive index, is the safe alternative for those that want to try the beautiful and difficult diatoms.

The medium rating.- CPG, AG, FG, FGL, and GG, although each have some good qualities they are not top quality, especially because of their poor drying behaviour. If there are better mountants, why use them? But try them with your subjects before you leave them aside.

The best media. Of course there are the remaining 7 media and can be split in two clear cut groups:

1) FMM, GAF, and even APT are cheap aqueous media, easy to buy or to make, dry well and useful for most of the subjects you can work with.

FMM (in the Karo, or Larry Legg's version) is probably the best general medium for young amateurs' work, and even for many professional jobs, being highly versatile and having many of the qualities of Glycerin, as well as preserving the green colour of chlorophyll for a while (if you fix your vegetable tissues with an adequate fixative). A 60% water solution of the commercial syrup is the more easy to use and best drying medium. I do not know how good the permanence of Larry Legg's fructose medium is, but Karo evidently has a very good antiseptic (safe, because it could be eaten).

GAF and APT, have many good qualities and have the same applications as FMM, but they need to be manually prepared, and stored with an antiseptic, and are not better than FMM.

2) NPM, GLG, PVA-L, PVA-G, is a special group of media that imparts to the mounted subjects a high transparency. And all of them have very good and most similar qualities, including very good optical and drying properties.

NPM is the most expensive, but the easiest to find and buy. For the others you need to get lactic acid (which seems to be not so difficult to obtain) and PVA (Polyvinyl Alcohol). If you can buy the professional grade PVA, it gives you the capacity to adjust the density of your medium. In my experience, to get PVA commercial glues seems not to be a problem. In only two searches (one in NY and the other here at Durango) I found two commercial transparent PVA paper glues. Not to be confused with the white glue that is also sold as PVA (Polyvinyl Acetate). I think PVA-G that does not need lactic acid, is also the best option for mounting some little creatures in which I am very interested, like tardigrades, rotifers and gastrotrichs, having the advantages of glycerin Jellies and none of its defects.

PVA-L. and GLG are the best alternatives to the much used but dangerous Hoyer's medium, with many advantages for the PVA-L formulation. When using them, if you make the formulas with PVA commercial glues, take care to replenish the media under the coverslip at frequent intervals until they stabilize. Then seal with care.

My mounting media favourites

Of course I use the AW method, and also use glycerin (PG) either as a mounting medium or as a clearing agent, or as an intermediate medium. I use Karo as a 'Jack of all trades', and NPM as my resinous mountant (well...when I remember that I am a grown-up, I also indulge myself with some special mountings with Damar) having at hand two additional little bottles of PVA-L, and PVA-G, made with commercial paper glue.

There is no one subject including histological or botanical sections (Karo, PVA-G) that I couldn't mount. (....except diatoms!) And I no longer have the problems of dehydrating, using solvents, heating slides and cover slides, using antiseptics, or slipping coverslides, or need stoves to dry slides.

And I even could use the Karo and the PVA-G formulas in the same way as the glycerin for step to step concentration when I try to mount my mixed collections of microinvertebrates (see the discussion of AW in the first part of this series, or the similar discussion on mounting GALA fixed materials in the first article on safe fixatives).

This is my advice for a cheap and safe set of mounting media, but which you select for your own favourites, is up to you. Experiment and select the mounting media that is best for the subjects in which you are most interested. You have 20 choices at your disposal.

NOTES:

I prepared most of the investigated media more than eight months ago. I used Listerine as my preferred antiseptic. None of the Listerine treated formulas had developed fungal infections but the Borax Jelly does after they melted in summer temperatures.

I left half the von Apáthy sucrose formula, and half the Fisher Borax Glycerin Jelly with double borax without adding Listerine to either. Both of them are now heavily contaminated. And I assure you that the von Apáthy smell is not good. Not in my country and possibly not in Europe, but G. Couger tells me that the best antiseptic for the Jellies, Phenic Acid (Phenol), is easy to obtain in the United States. Add 1% w/w after finishing the preparation of the medium.

REFERENCES & GRATITUDES

Howard Webb
MICSCAPE MAGAZINE, November 2001

J.M. Cavanihac
Montage des lames,
http://www.microscopies.com/

Brian Adams
MICSCAPE MAGAZINE,
February 1999

Jean Legrand
Construction d'une hotte de sechage,
www.microscopies.com/DOSSIERS/Magazine/Articles/JLegrand-Hotte/Etuve.htm

Richard Howey, David Walker, & Mol Smith for their help and assistance.

Equipment used to create images for this work:
National Optical DC3-163P microscope, with an integrated 0.3 Mpixel Motic DC·3 digital camera, with automatic white balance and controlled by its own capture software. The microscope was equipped with DIN planachromatics 4x (NA 0.10), 10x (0.25), 40x (0.65) and 100x HI (1.25) objectives, an Abbe condenser NA 1.25 and a 12V, 20W lamp included in the base and capable of critical illumination. The camera later used was a 2 Mpixel Logitech 9000 Quickcam Pro.

By Walter Dioni. Originally published in Micscape Magazine.
SAFE MICROSCOPIC TECHNIQUES
FOR AMATEURS.
MOUNTING MICROSCOPIC SUBJECTS

The full suite of original articles by Walter Dioni can be found at the link below. Many more are at this link than those included in this volume.
http://www.microscopy-uk.org.uk/mag/wd-articles.html

Printed in Great Britain
by Amazon.co.uk, Ltd.,
Marston Gate.